This book shows how modern cosmology and astronomy have led to the need to introduce dark matter in the universe. Some of this dark matter is the familiar form of protons, electrons and neutrons, but most of it must have a more exotic form. The favoured, but not the only, possibility is neutrinos of non-zero rest mass, pair-created in the hot big bang and surviving to the present day. After a review of modern cosmology, this book gives a detailed account of the author's recent theory in which these neutrinos decay into photons which are the main ionising agents for hydrogen and nitrogen in the interstellar and intergalactic medium. This theory, though speculative, explains a number of rather different puzzling phenomena in astronomy and cosmology in a unified way and predicts values of various important quantities such as the mass of the decaying neutrino and the Hubble constant. Written by a cosmologist of the first rank, this topical book will be essential reading to all cosmologists and astrophysicists.

CAMBRIDGE LECTURE NOTES IN PHYSICS

1. Clarke: The Analysis of Space-Time Singularities
2. Dorey: Exact S-Matrices in Two Dimensional Quantum Field Theory
3. Sciama: Modern Cosmology and the Dark Matter Problem
4. Veltman: Diagrammatica – The Path to Feynman Rules

CAMBRIDGE LECTURE NOTES IN PHYSICS 3

General Editors: P. Goddard, J. Yeomans

Modern Cosmology and the Dark Matter Problem

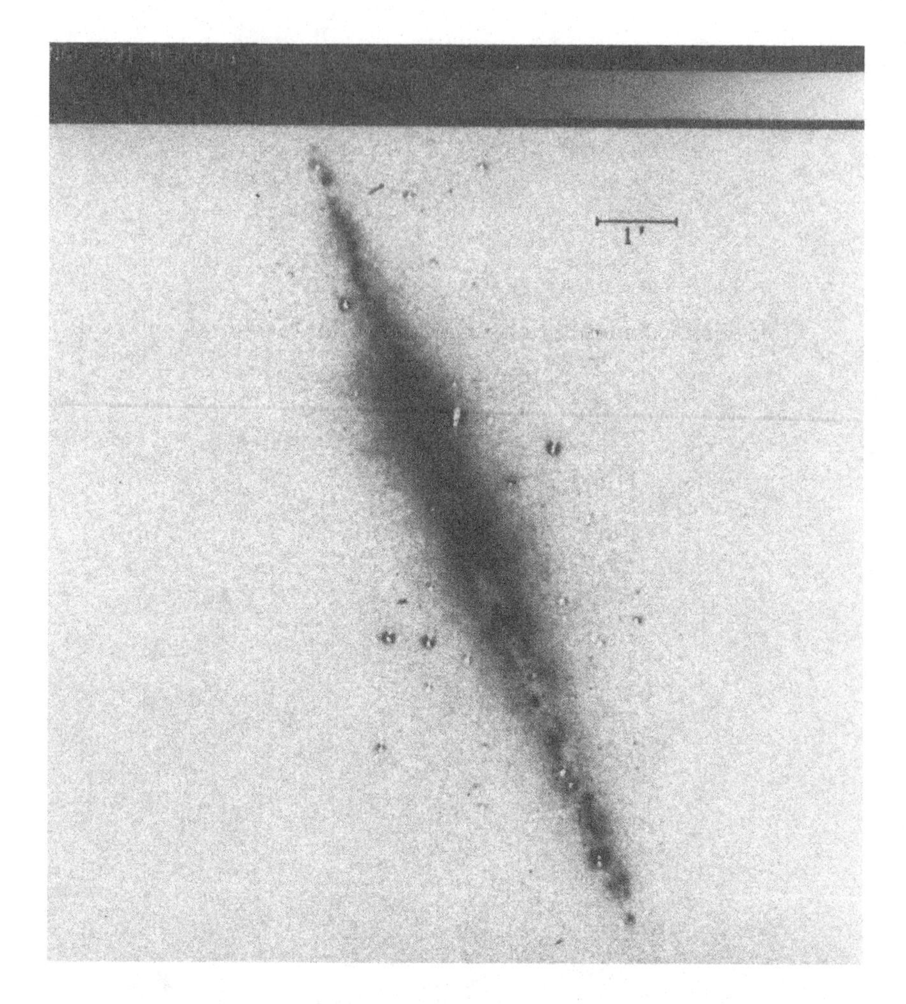

This Hα picture of the edge-on spiral galaxy NGC 891 reveals a widespread distribution of ionised gas stretching several kioloparsecs from the central plane of the galaxy. The origin of this ionisation is not understood, but it could be due to ultraviolet photons emitted by dark matter neutrinos in the halo of the galaxy. [Photograph by courtesy of R. J. Rand, from Rand, Kulkarni and Hester 1990].

Modern Cosmology and the Dark Matter Problem

D. W. SCIAMA
SISSA and ICTP, Trieste
Department of Physics, Oxford
Fellow of Churchill College, Cambridge

CAMBRIDGE
UNIVERSITY PRESS

Published by the Press Syndicate of the University of Cambridge
The Pitt Building, Trumpington Street, Cambridge CB2 1RP
40 West 20th Street, New York, NY 10011–4211, USA
10 Stamford Road, Oakleigh, Melbourne 3166, Australia

First published 1993
Reprinted with new section 'Recent Developments' 1995

A catalogue record for this book is available from the British Library

Library of Congress cataloguing in publication data applied for

ISBN 0 521 43848 9 paperback

Transferred to digital printing 2003

To Lidia, Susan and Sonia

Contents

Preface

I started writing *Modern Cosmology* in 1969, just four years after the discovery of the 3 K cosmic microwave background. The significance of that remarkable discovery was rapidly appreciated by cosmologists, and it naturally dominated a large part of my book. Now, nearly a quarter of a century later, a new topic has come to dominate cosmology, namely, the dark matter problem. This problem, however, is not at all well understood. According to modern estimates some of the dark matter is in the form of ordinary particles — protons, neutrons and electrons — while some of it has a more exotic character. We do not know what form the ordinary dark matter takes, and we do not even know the identity of the exotic dark matter. Yet together they are a pervasive and indeed dominating constituent of the universe, in galaxies, groups and clusters of galaxies and intergalactic space. I therefore thought it desirable to update my book by writing a connected account of what has now become the single most important problem in astronomy and cosmology.

I must confess that I have a second reason for writing this book. In 1990 I proposed the idea that most of the widespread ionisation of hydrogen observed in our Galaxy is produced by photons emitted by decaying dark matter neutrinos of non-zero rest-mass. The original motivation for this proposal was that the observed ionisation was puzzling astronomers because it seemed difficult to account for in terms of known sources of ionisation.

This theory, once proposed, rapidly took on a life of its own. Because neutrinos are ubiquitous, their ionising photons would have a major impact on a large variety of phenomena in astronomy and cosmology. In addition, a non-standard theory of particle physics would be needed to provide the required decay lifetime (~ 2 to 3×10^{23} secs), so that a new link would be created between observational astronomy and elementary particle physics.

An unexpected consequence of the theory is that its domain of validity is non-zero but extremely small. By a remarkable numerical quirk, the (monochromatic) energy of the decay photons, the rest-mass of the decaying neutrinos, and the Hubble constant are all observationally constrained to have values specified with a precision of about 1 percent. This means that the theory is strongly predictive (and vulnerable). It also means that either the theory is correct or we are faced with a remarkable set of chance coincidences.

Because so many phenomena of astronomy, cosmology and elementary particle physics become linked together by this decaying neutrino hypothesis, I thought it desirable to write a connected account of the hypothesis and its observational and theoretical implications. My aim has been more to convey the basic ideas involved than to write a scholarly and systematic account of all these implications. Indeed such an account would require a book far longer than the present one, and it would also require a far greater expertise than this author possesses. I hope, therefore, that the reader will forgive the sketchy nature of my treatment. I have attempted to compensate for this by giving sufficient references to permit a deeper study of any particular topic of interest. I have also given a somewhat leisurely account of cosmological models in section 3.2. Although the style of this section is out of keeping with the rest of the book I have adopted it deliberately because an understanding of cosmological models and their implications is fundamental to the dark matter problem.

It would not be practicable to acknowledge all the scientists who have helped me with discussions and advice. I shall therefore restrict myself to thanking those who have been co-authors with me of papers concerned with the neutrino decay hypothesis or who have commented on draft sections of this book. They are: D. Burstein, L. Buson, A. C. Fabian, F. Gabbiani, A. Lanza, A. Masiero, A. L. Melott, P. Maloney, T. Naylor, M. Persic, S. T. Petcov, M. J. Rees, Y. Rephaeli, P. Salucci, C. Sarazin, D. Scott and R. F. Stark. I am also grateful to my colleagues who have worked so devotedly on our proposed satellite experiment described in Chapter 13, and in particular, C. S. Bowyer, A. Gimenez, C. Morales, and R. Stalio. Finally, I want to acknowledge the astronomer to whom I owe the most, Ron Reynolds, many of whose perceptive papers on the ionisation state of our Galaxy and of the outside universe have inspired my neutrino decay theory.

I am also grateful to those who have helped me to prepare this book. Rufus Neal of the Cambridge University Press has been a tower of strength as my editor, Antonio Lanza has spent many hours working on the technical aspects of dealing with LaTeX, John Miller and Paolo Salucci have helped with production problems and Brigitte Salucci has tirelessly turned a messy manuscript into a publishable text. The Oxford University Physics Photographic Unit made beautiful prints of all the figures. In addition the Italian Ministry of Universities and Scientific and Technological Research has provided financial support.

My wife Lidia and my daughters Susan and Sonia have been a source of great joy, and have made it possible for me to write this book. I dedicate it to them with all my love.

Recent Developments

The publication of this reprint enables me to describe the main observational developments in the dark matter problem which have occurred in the last two years. The most significant development concerns the observational search for "machos" in the halo of our Galaxy. These machos are to be contrasted with the wimps discussed on page 63. They are massive compact halo objects such as black holes, neutron stars, brown dwarfs, or jupiters (e.g. Carr 1994).

Paczynski pointed out in 1986 that such objects might be detected when they pass in front of background stars, because their gravitational lens effect would lead to a transient brightening of these stars (microlensing, page 63). The probability of a detectable event occurring at any time for a particular star is very low, about 10^{-6}, so that a vast number of stars must be monitored. Three groups are conducting a search for such events, named EROS, MACHO and OGLE, and all three have reported the detection of candidate events towards the Large Magellanic Cloud and the galactic bulge.

These events have given rise to considerable discussion. In a recent analysis Paczynski (1995) writes as follows: "The results available so far neither prove nor disprove the hypothesis that dark matter is made of machos. All events detected so far are compatible with the lenses being ordinary stars. My personal conclusion is that machos, if they exist, will be detected within the next few years".

The remaining observational developments which I shall describe have implications for the decaying neutrino theory which occupies Part Three of the book. These developments are the following:

1) Hα observations of the intergalactic HI cloud 1225+01 by

Vogel *et al.* (1995).

2) The Hubble Space Telescope observation of Cepheid variables in M100, a galaxy in the Virgo cluster, by Freedman *et al.* (1994).

3) Millimetre observations of the C^0/CO abundance ratio in the molecular cloud G34.3+0.2 by Little *et al.* (1994).

4) The probable HST observation of the Gunn–Peterson effect in HeII in the spectrum of the quasar 0302-003 by Jakobsen *et al.* (1994).

5) Confirmation of the COBE observations of the anisotropy $\Delta T/T$ in the cosmic microwave background, and its extension to smaller angular scales, by various groups.

I now describe these developments and their implications for the decaying neutrino theory.

1) Vogel *et al.* 's failure to observe an $H\alpha$ line from 1225+01 led them to impose the following conservative constraint on the intergalactic hydrogen-ionising flux $F(0)$ at zero red shift:

$$F(0) < 1.5 \times 10^5 \text{ photons cm}^{-2} \text{ sec}^{-1}.$$

This new upper limit is 4 times smaller than the one which we adopted on page 109. As explained on page 163, it would lead to a more restrictive upper limit on the monochromatic energy E_γ of a decay photon,

$$E_\gamma < 13.8 \text{ eV},$$

instead of 14.6 eV. Thus we would now derive

$$E_\gamma = 13.7 \pm 0.1 \text{ eV}.$$

This has an immediate implication for the mass m_{ν_1} of the decaying neutrino. If the mass m_{ν_2} of the secondary neutrino in the decay can be neglected ($m_{\nu_2}^2 \ll m_{\nu_1}^2$), we would have $m_{\nu_1} = 2E_\gamma$ (page 119) and so

$$m_{\nu_1} = 27.4 \pm 0.2 \text{ eV}.$$

Thus E_γ and m_{ν_1} would be determined with a precision of 1 percent.

The cosmological density parameter Ω would remain close to 1 (page 145), and if it is exactly 1 we would now have for the Hubble constant H_0

$$H_0 = 54 \pm 0.5 \text{ km sec}^{-1} \text{ Mpc}^{-1},$$

instead of 56.3 (page 146), if the cosmological constant is zero.

The substantial reduction in the upper limit on $F(0)$ would be compatible with our suggestion (page 163) that the cosmological decay photon flux $F_{ext}(z)$ at red shift z is mainly responsible for the ionisation of hydrogen in the intergalactic medium and in Lyman α clouds for the observed range of red shifts from zero to five. For example, since $F_{ext}(z) = (1+z)^{3/2} F_{ext}(0)$ (page 163), we would now have

$$F_{ext}(3) < 1.2 \times 10^6 \text{ cm}^{-2} \text{ sec}^{-1}.$$

The actual hydrogen-ionising flux $F(z)$ at $z = 3$ is constrained by the proximity effect for quasars (page 103) as follows:

$$F(3) \geq 5 \times 10^5 \text{ cm}^{-2} \text{ sec}^{-1}.$$

The contribution of quasars to $F(0)$ is much less than the new upper limit of 1.5×10^5 cm^{-2} sec^{-1}. Their contribution to $F(3)$ (page 113) is still controversial, but could be a few times less than the lower limit of 5×10^5 cm^{-2} sec^{-1}. Thus if E_γ is fairly close to its new upper limit of 13.8 eV we could still have decay photons mainly responsible for the hydrogen–ionising flux at $z = 0$ to 5. Evidence to the contrary has recently been obtained by Williger *et al.* (1994) who find, for one quasar, that $F(4.2)$ is only one third of $F(3)$. However, as these authors point out, this result needs confirmation from observations of other quasars.

The new, more stringent, constraint on $F(z)$ would resolve the problem for the decaying neutrino theory concerning a possible excess column density N(HeI) implied by it for Lyman α clouds (pages 166-8). The reason is that N(HeI)/N(HI) in the clouds is proportional to the ionisation rate of HI, and this rate must now be substantially reduced.

A further consequence of the new upper bound on E_γ is that we must now withdraw our claim (page 89) that decay photons are mainly responsible for the ionisation of interstellar nitrogen, whose ionisation potential is 14.5 eV. As discussed in detail elsewhere (Sciama 1995), one can attribute this ionisation to photons emitted by O stars which then leak through the interstellar system of opaque HI clouds. The required flux of these photons is only about 1 percent of that invoked by Miller and Cox (1993), Domgorgen and Mathis (1994) and Dove and Shull (1994), who have attempted to attribute all the hydrogen–ionisation of the interstellar medium to photons from O stars which take advantage of the porosity of the cloud system.

2) Freedman *et al.* 's observation of Cepheid variables in M100, a galaxy in the Virgo cluster, led them to derive a Hubble constant H_0 of 80 ± 17 km sec^{-1} Mpc^{-1}. This is similar to the value 87 ± 7 km sec^{-1} Mpc^{-1} obtained by Pierce *et al.* (1994) from Cepheids observed with a ground-based telescope in another Virgo cluster galaxy NGC4571. These values are significantly greater than our prediction of 54 ± 0.5 km sec^{-1} Mpc^{-1}. One cannot, however, obtain a reliable result for H_0 from observations of Cepheids in only one or two galaxies of the Virgo cluster because this cluster is spread out over a significant distance along the line of sight (e.g. Young and Currie 1995), and one cannot tell where an individual galaxy is located relative to the cluster centre. Fortunately Cepheids in several more galaxies in the cluster should be observed by HST during 1995, which may enable this uncertainty to be reduced, if suitable allowance can be made for selection effects.

In this connexion it is interesting to note that methods for determing H_0 based on supernovae of type Ia (Riess *et al.* 1995, Saha *et al.* 1995) and direct, but model–dependent, methods using the Sunyaev–Zeldovich effect and gravitationally lensed quasars, tend to give lower values, in the range 50 – 70 km sec^{-1} Mpc^{-1}, and therefore closer to our prediction. Such lower values would be more compatible with estimates of the age of the universe (if the cosmological constant is zero) than is a value in the 80's (page 60). The resolution of this problem would be of major importance for cosmology generally, as well as for the decaying neutrino theory.

3) The observations and modelling of the C^0/CO abundance ratio in the molecular cloud G34.3+0.2 by Little *et al.* (1994) have lent some support to Tarafdar's suggestion (pages 140-2) that the anomalously large value of this ratio in molecular clouds is mainly due to the photodissociation of CO by decay photons produced in the clouds. According to Little *et al.* this suggestion would agree with their results, but only if one adopts for the photodissociation cross–section of CO a value close to Tarafdar's original choice of 7.8×10^{-17} cm^2, rather than the revised value of 10^{-17} cm^2 which we used on page 141. The reason for the revision was that we then took $E_\gamma = 14.6$ eV, whereas Tarafdar assumed that $E_\gamma = 13.8$ eV. This relatively small change in the adopted photon energy implied a large change in the cross–section because this energy

range lies below the photodissociation continuum of CO and the main process involved is line absorption into predissociated bound states (page 140).

If we now accept that E_γ is actually less than 13.8 eV, we must reconsider the value of the cross-section which should be adopted. The absorption line spectrum of CO in the relevant energy range has been measured by Letzelter *et al.* (1987). If we require that the cross-section should lie within a factor 2 of Tarafdar's choice, we find from the spectrum of Letzelter et al, that the photon energy E_γ must lie in the range

$$E_\gamma = 13.65 \ - \ 13.73 \text{ eV},$$

and so m_{ν_1} must lie in the range

$$m_{\nu_1} = 27.30 \ - \ 27.46 \text{ eV},$$

(if $m_{\nu_2}^2 \ll m_{\nu_1}^2$). If these astronomical arguments are correct, E_γ and m_{ν_1} would be determined with even greater precision than in 1) above.

4) An important step forward in our understanding of the intergalactic medium and of the spectrum of the incident ionising flux has been taken with the discovery by Jakobsen *et al.* (1994) of a probable substantial HeII Gunn–Peterson effect in the spectrum of the quasar Q0302-003 (cf. pages 100-2 for the HI and HeI Gunn-Peterson effects). One might be able to use these effects to test the decaying neutrino theory by determining from them the spectral emissivity ratio $\Phi_{\rm HI}/\Phi_{\rm HeII}$ for the sources of the ionising flux, after allowing for various absorption processes (Madau and Meiksin 1995). If the resulting ratio were too steep for quasars, this could fit in with the decaying neutrino theory, whose value for $\Phi_{\rm HI}$ may exceed that of quasars by a factor of a few, but whose photons cannot ionise HeII. More observational data are needed before this test could be carried out.

5) There have now been a number of confirmations and extensions to smaller angular scales of the COBE discovery of anisotropies $\Delta T/T$ in the cosmic microwave background (page 176). Observations of this kind should soon be able to test the prediction, discusssed in chapter 12, that the early reionisation of the universe due to decay photons is suppressing $\Delta T/T$ on sufficiently small angular scales (that is, on scales smaller than the angle subtended by the horizon at the last scattering surface). In

particular, the so-called Doppler peaks should be suppressed (e.g. White *et al.* 1994). Another possible observational test of early reionisation arises from the resulting increase by a factor of order 10 in the linear polarisation of the microwave background, to a value of order $0.1\Delta T/T$.

Finally, we note that the space experiment EURD, designed to search for the decay line from neutrinos near the sun using a Spanish minisatellite (pages 189-192) is now due to be launched by a Pegasus rocket in February 1996. This experiment should still provide the most decisive test of the decaying neutrino theory.

D.W. Sciama
February 1995

REFERENCES

Carr, B., 1994, *Ann. Rev. Astr. Ap.*, **32**, 531.
Domgorgen, H. and Mathis, J.S., 1994, *Ap. J.*, **428**, 647.
Dove, J.B. and Shull, J.M., 1994, *Ap. J.*, **430**, 222.
Freedman, W.L. et al, 1994, *Nature*, **371**, 757.
Jakobsen, P. et al., 1994, *Nature*, **370**, 35.
Letzelter, C. et al., 1987, *J. Chem. Phys.*, **114**, 273.
Little, L.T. et al. 1994, *M.N.R.A.S.*, **271**, 649.
Madau, P. and Meiksin, A., 1995, *Ap. J.*, in press.
Miller, W.W. and Cox, D.P., 1993, *Ap. J.*, **417**, 579.
Paczynski, B., 1995, in Proc. 5th Astrophys. Conf. Maryland.
Pierce, M.J. et al., 1994, *Nature*, **371**, 385.
Riess, A.G. et al., 1995, *Ap. J.*, **438**, L17.
Saha, A. et al., 1995, *Ap. J.*, **438**, 1.
Sciama, D.W., 1995, *Ap. J.*, in press.
Vogel, S.N. et al., 1995, *Ap. J.*, March 1 issue.
White, M. et al., 1994, *Ann. Rev. Astr. Ap.*, **32**, 319.
Williger, G.M. et al. 1994, *Ap. J.*, **428**, 574.
Young, C.K. and Currie, M.J., 1995, *M.N.R.A.S.*, in press.

Part One
Dark Matter in Astronomy and Cosmology

1
Dark Matter in Galaxies

1.1 Introduction

The detection of dark matter in astronomy has a long history. In past years it was called "the astronomy of the invisible". The story begins in 1844 when, by chance, two different dark matter problems were identified. In that year it was noted that the planet Uranus had moved away from its calculated position by as much as two minutes of arc. In the same year F. W. Bessell drew attention to the sinuous motion of the star Sirius, the brightest star in the sky.

The subsequent development of the Uranus problem led to one of the most famous stories in the history of astronomy. In 1845 J. C. Adams, who had just ceased to be an undergraduate at Cambridge University, succeeded in calculating fairly accurately the position of a hypothetical planet whose gravitational effect on Uranus might be responsible for its disturbed motion. He attempted unsuccessfully to interest the Astronomer Royal G. B. Airy in this prediction. Apparently Airy had attributed the discrepancy to a departure from Newton's law of gravity. Perhaps also he was unimpressed by the student's youth.

Independently of Adams, in 1846 the Frenchman Le Verrier calculated the position of the hypothetical planet with a precision of 1 degree. (For a much shortened version of the needed calculations see Lyttleton 1958). Le Verrier contacted a German astronomer, Galle, at the Berlin Observatory, who rapidly succeeded in observing a new planet (Neptune) within 1 degree of the predicted position. The discovery of this planet (no longer "dark") is widely considered to be a triumph of nineteenth century science, and naturally became the subject of chauvinistic controversy. However, today Adams and Le Verrier are accorded equal honour by astronomers.

Meanwhile Bessell's suggestion of 1844 that Sirius was being disturbed by the gravitational action of a faint companion star was also followed up. In 1862 a telescope maker Alvan Clarke observed this faint companion (Sirius B). It later became an important object for astrophysics, when in 1915 it was found to be white in colour. This meant that it was hot, and therefore a dwarf. The important class of white dwarfs had been discovered.

1.2 Dark Matter near the Sun

The first suggestion of a possible widespread distribution of dark gravitating matter in our Galaxy seems to have been made by Kapteyn (1922) and Jeans (1922), followed by Lindblad (1926). They attempted to use the observed motions of nearby stars at right angles to the plane of the Galaxy to derive the total density of matter near the sun. Kapteyn wrote in the abstract of his paper "...it may be possible to determine *the amount of dark matter* from its gravitational effect". He concluded that "this mass [of dark matter] cannot be excessive."

The first claim that there exists a substantial amount of dark matter near the sun was made by Oort (1932, 1965), who also used the observed vertical motions of stars. He derived for the total density of matter the value $\sim 0.2\ M_\odot\ pc^{-3}$, which is about twice as large as the observed density of matter near the sun.

Unfortunately today, seventy years later, the question of the amount (if any) of dark matter near the sun is still controversial. There is no agreement about the choice of the most suitable tracer population of stars, or about the various assumptions which have to be made before the governing equations can be solved. Moreover, systematic errors turn out to be important.

To illustrate the existing differences of opinion we quote from two recent studies, which give references to earlier discussions in the literature. Kuijken and Gilmore (1991) (KG) conclude that "there remains no significant evidence for any unidentified matter associated with the galactic disk near the sun". By contrast, Bahcall, Flynn and Gould (1992) (BFG) find that "a model with no dark matter is inconsistent with the data at the 86% confidence level."

This dispute turns out to be important for the theory of decaying dark matter neutrinos discussed later in this book (the DDM theory). As we shall see in the next section, in order to account for the observed flat rotation curve of the Galaxy, one needs to introduce a dark halo whose total column density at the sun and at right angles to the galactic plane is about 220 $M_{\odot}$ pc^{-2} (Caldwell and Ostriker 1981). To determine the volume density of this halo at the sun we need to know its scale height. (In this book we will mean by scale height the column density on one side of the galactic plane divided by the volume density in the plane). We shall see in Chapter 9 that in the DDM theory the scale height $\sim$ 3 kpc, corresponding to a flattened halo. We must make sure that this required flattening does not lead to local densities so high as to violate any constraints imposed by the Oort-type analysis.

Two constraints are important here, namely those involving the halo volume density at the sun, and the column density out to $\sim$ 1 kpc which is the present limiting height of the Oort-type analysis. The DDM values for these quantities are $\sim$ 0.03 $M_{\odot}$ pc^{-3} and $\sim$ 37 $M_{\odot}$ pc^{-2} respectively. According to Gilmore (1991) these values are compatible with both KG and BFG when allowance is made for all the uncertainties. In fact it is possible that essentially all the dark matter near the sun is associated with the halo (Binney, May and Ostriker 1987), although in all Oort-type analyses to date the dark matter is assumed to have a scale height less than 1 kpc.

1.3 Dark Matter in Galactic Halos

Perhaps the first suggestion of dark matter in galactic halos was made by Freeman in 1970. He pointed out that the rotation curves of NGC 300 and M33, measured in the 21 cm line of neutral hydrogen, did not show the expected Keplerian decline beyond the optical radii of these galaxies. He concluded that "there must be in these galaxies additional matter which is undetected...Its mass must be at least as large as the mass of the detected galaxy, and its distribution must be quite different from the exponential distribution which holds for the optical galaxy." Similar results were found by later workers. In addition Ostriker and Peebles (1973) suggested that spiral galaxies need dark halos to stabilise them against bar-like instabilities.

This work was then summarised in two influential review articles (Faber and Gallagher 1979, Rubin 1979) which proposed that normal spiral galaxies contain substantial amounts of dark matter located at great distances from the central regions, and so distributed as to lead to an essentially constant rotation velocity outside these central regions. While the shape of the dark matter halos could not be determined from these considerations, it was clear that if a halo were spherical the mass within radius r would have to increase like r to produce a rotation velocity independent of r. Equivalently, the density in such a halo would have to decrease like $1/r^2$.

In the succeeding years many studies were made of this phenomenon. (For a recent review of this and other dark matter problems see Ashman 1992). Today the flatness of rotation curves is generally accepted (although it should be stressed that these curves are not literally flat, and that observed small deviations from flatness are themselves used to make important deductions (e.g. Persic and Salucci 1988, 1990 a, b)). Two surveys of this question written in 1991 give an up to date account of the situation (Persic and Salucci 1992b, Rubin 1991). To specify how far out the rotation curve remains flat it is convenient to follow the discussion of Freeman (1970) and to work in terms of the optical scale length of each galaxy. Freeman pointed out that the observed radial distribution of surface brightness of the disks of normal spiral galaxies is exponential

$$I(r) = I_0 e^{-r/R_D},$$

where R_D is a constant for each galaxy (its optical scale length). Even allowing for the contributions of stellar bulges and interstellar gas, one would expect to reach the Keplerian regime ($v \propto r^{-1/2}$) at $r \sim 6r_D$, unless dark matter is present in the outer regions. However, in NGC 3198, for example, ("everyone's favourite"), the rotation curve observed in HI at 21 cm is flat out to 11 R_D (van Albada *et al.* 1985). This rotation curve is shown in Fig. 1.1.

It is still not possible to construct a unique model for the way in which stars, gas and dark matter combine to give the observed rotation curve of a spiral galaxy. Despite this, an important dis-

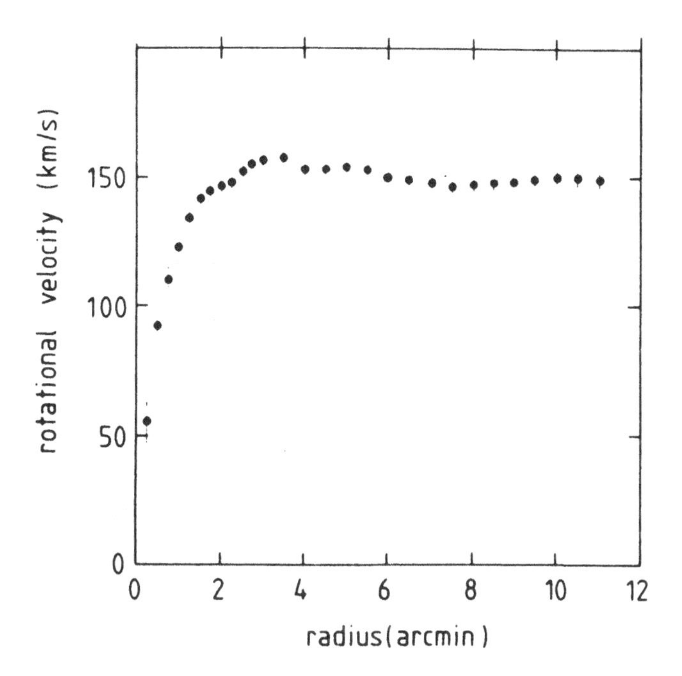

Fig. 1.1 The rotation curve of NGC 3198 from 2.68 to 29.5 kpc (for an assumed distance of 9.2 Mpc) [From K. Begeman 1987].

covery has been made by Persic and Salucci (1988, 1990a, b) concerning the relative total amounts of dark and visible matter in spiral galaxies of different luminosity. By a careful study of the small velocity gradients in the rotation curves they were able to show that

$$\frac{M_{halo}}{M_{disk}} \propto L_B^{-0.7},$$

where L_B is the disk luminosity in the blue.

One galaxy which we can study in more detail is, of course, our own, since we possess very detailed information for it about the distribution of stars and gas. Nevertheless, it is not possible to derive a unique model for the distribution of dark matter needed to account for the observed rotation curve of the Galaxy (Fig. 1.2).

An influential model was proposed by Caldwell and Ostriker

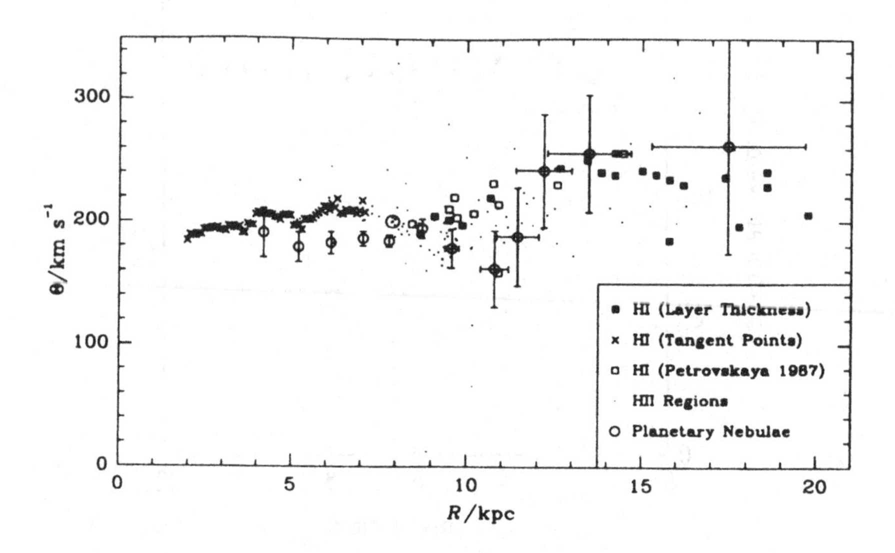

Fig. 1.2 The rotation curve of our Galaxy [From M. R. Merrifield 1992: Reproduced courtesy of *The Astronomical Journal*].

(1981) who used all the information available to them concerning the distribution of stars and gas in the Galaxy and, of course, the rotation curve. They parametrised the density ρ_d of dark matter in the plane by a core-halo type model of the form

$$\rho_d = \frac{\rho_0}{1 + r^2/a^2},$$

and assumed that the halo is spherical.

They found a good fit to the data with

$$\rho_0 = 1.37 \times 10^{-2}\ \mathrm{M_\odot\ pc^{-3}},$$

$$a = 7.8\ \mathrm{kpc}.$$

Caldwell and Ostriker also derived the total column density Σ_d of dark matter at right angles to the Galactic plane as a function of distance from the centre of the Galaxy. They obtained for Σ_d

at the sun's position the value

$$\Sigma_{d,\odot} \sim 220 \text{ M}_\odot \text{ pc}^{-2}.$$

A reasonable estimate for the uncertainty in this quantity would be 30 per cent. We shall find in chapter 9 that $\Sigma_{d,\odot}$ plays a key role in our decaying dark matter theory and that, in conjunction with $H\alpha$ data for the Galaxy, it will enable us to derive in that theory a value for the radiative lifetime of decaying neutrinos.

We now consider the likely shape of the dark halos around galaxies. A useful survey of this question has been given by Ashman (1992). N-body simulations of the dissipationless formation of halos (Frenk *et al.* 1988, Katz 1991, Katz and Gunn 1991, Dubinski and Carlberg 1991, Dubinski 1992, Warren *et al.* 1992) indicate that the likely final shape is flattened, and even triaxial. The flattest halos obtained in the absence of dissipation had an axial ratio $q \sim 0.4$.

A direct observational attack on the problem for our Galaxy has been made by van der Marel (1991). He studied the metal-poor stellar halo and showed from Jeans' stellar hydrodynamical equation that its axial ratio and vertical velocity dispersion put a lower limit on q. He found in this way that $q > 0.34$. A study has also been made (Haud 1992) of the dynamical influence of a triaxial dark halo on the outer arm high velocity cloud of our Galaxy. In addition, one can use the observed shapes of galaxies to draw plausible conclusions about the shapes of their dark halos (Lambas, Maddox and Loveday 1992).

Recently it has been pointed out (Kuijken and Tremaine 1993) that the shapes of orbits in the galactic disk are influenced by the halo potential: in particular if the disk lies in a principal plane of a triaxial halo the closed orbits in the disk would be elliptical. Kuijken and Tremaine made a detailed study of various kinematic data for our Galaxy and concluded that its disk indeed has an elliptical distortion such as might be caused by a triaxial halo, although more general distortions would also be compatible with the data.

The method that has provided the best evidence for flattened halos uses polar ring galaxies. These are galaxies which have rings of gas, dust and stars in rotation on orbits nearly perpendicular to the plane of the central disks. By studying the rotation curves in the two planes for the polar ring galaxy NGC 4650A, Sackett and Sparke (1990) and Sackett (1991) obtained for this galaxy $q \sim 0.4$.

A number of interesting dynamical consequences can be drawn if flattened dark halos are common. For example, they could be responsible for the warps usually seen in the outer HI disks of spiral galaxies (Sparke and Casertano 1988, Kuijken 1991). A full theoretical treatment of this problem would be difficult, as one would have to treat the halo as a non-rigid object (Binney 1992). Flattened halos might also explain the observed fact that the orientation of rings with respect to their central disks is non-random: too many polar rings are too nearly polar (Schweizer *et al.* 1983). This would follow because a ring with an intermediate inclination would be dynamically unstable in the presence of a flattened halo. Another interesting question concerns the formation of galaxies in asymmetric dark halos (Subramanian 1988).

The relation of a flattened dark halo in our Galaxy to the Oort problem of the dark matter density near the sun has already been discussed. It would be desirable to reanalyse the Oort problem assuming that the dark matter has a scale height exceeding 1 kpc.

The most important question raised by the discovery of dark matter in galaxies is the elucidation of its nature. We defer consideration of this question to chapter 4, after we have discussed the dark matter in clusters of galaxies and in the universe as a whole.

2
Dark Matter in Clusters of Galaxies

2.1 Introduction

Extragalactic dark matter was discovered by Zwicky (1933) only a year after Oort's original study of dark matter near the sun. Zwicky showed that the velocity dispersion of galaxies in a cluster far exceeds that expected for a gravitationally bound stationary system if the only contribution to the gravitational field of the cluster comes from the galaxies themselves. Useful reviews of this dark matter problem can be found in Faber and Gallagher (1979), Rood (1982), Sarazin (1988) and Oegerle *et al.* (1990).

The first detailed analyses of the problem were based on the assumption that the dark matter in the cluster is distributed in the same way as the visible matter - the " mass follows light" assumption. For example, Kent and Gunn (1982) collected optical red shifts for about 300 galaxies in the Coma cluster and by using this assumption found that the total mass of the cluster was rather well determined. Their results depend on the distance of Coma, and so on the value of the Hubble constant H_0. Throughout this book we shall use the usual parametrization for H_0, namely

$$H_0 = 100\ h \text{ km sec}^{-1} \text{ Mpc}^{-1},$$

with the observational uncertainty in its value given by

$$\frac{1}{2} \leq h \leq 1.$$

(see page 59). Then Kent and Gunn found that the total projected mass M of Coma within 3 degrees of its centre is

$$M = 5.8 \times 10^{15}\ h\ \mathrm{M}_{\odot}.$$

It has become conventional in astronomy to express the masses of luminous objects in terms of "mass to light ratios", where M/L for the sun is taken as unity, and L could be blue light, visible light etc. Kent and Gunn's result can then be written

$$\frac{M}{L_B} = 362\ h.$$

By contrast, the mass to light ratios found for the luminous portions of individual galaxies range over $M/L_B \sim (2 \text{ to } 24)\ h$, with the large values corresponding to the E and S0 galaxies which predominate in compact regular clusters. The extent of the dark matter problem in clusters is thereby clear. Whether the individual galaxies in a cluster are able to retain their own dark matter in the cluster environment would require a separate discussion.

These results would have to be changed if the "mass follows light" assumption were dropped (e.g. Bailey 1982, The and White 1986, Merritt 1987). In particular, if the dark matter is more concentrated towards the centre of the cluster than is the visible matter, then the total mass corresponding to a given velocity dispersion profile for the galaxies would be reduced (by at most a factor 5 according to Merritt). The optical data alone cannot resolve this ambiguity.

2.2 X-Ray Studies of Clusters

This situation changed in principle in 1971/2 when it was discovered that clusters of galaxies are powerful emitters of x-rays (this subject is surveyed in Sarazin 1988). It became evident that most of this emission is due to a hot intracluster gas ($T \sim 10^6$ to 10^8 K), whose high temperature corresponds to the gas being in hydrostatic equilibrium in the gravitational field of the cluster. Since the velocity distribution of the hot gas particles could be assumed to be isotropic (unlike that of the galaxies), analysis of the temperature and density distribution of the gas is a simple, and fairly model-independent, method of determining the distribution of the gravitating mass within clusters of galaxies. The first attempts to carry out such analyses (e.g. by Cowie, Henriksen and Mushotzky 1987, Hughes *et al.* 1988, The and White 1988) were hampered by the lack of spatially resolved x-ray spectra. At that time only a single x-ray spectrum integrated over most of the cluster was available. More recently it has become possible to measure both the temperature and density profiles of the gas in a cluster, and this has led to a direct determination of the distribution of gravitating mass in Perseus (Eyles *et al.* 1991), A85 and A2199 (Gerbal *et al.* 1992) and Coma (Briel *et al.* 1992).

To determine the distribution of dark matter one has to subtract out the contribution to the gravitational potential coming from the galaxies and the x-ray emitting gas. Eyles *et al.* found in this way that their data for Perseus strongly favour models in which the dark matter is more centrally concentrated than the galaxies or the gas. A similar result has been found for the Abell clusters A85 and A2199 by Gerbal *et al.* (1991), and for Coma by Briel *et al.* (1992).

More recently it has become possible to make a similar analysis for the cluster A665. This will be important later on when we come to discuss observational evidence relevant to our decaying neutrino hypothesis. The cluster A665 is interesting also because it is the richest cluster in the Abell catalogue, and because it has the best-established Sunyaev-Zeldovich effect (a decrease in the temperature of the cosmic microwave background resulting from Thomson scattering induced by the hot electrons in the cluster (Birkinshaw 1990)).

X-ray data for A665 have been obtained by Hughes and Tanaka (1992) using the Ginga satellite. The combined x-ray and optical data (Oegerle *et al.* 1991) have allowed them to construct a detailed self-consistent mass model of A665. This model includes:

1) a galaxy component of radial profile given by

$$\rho_G \sim \left[1 + \left(\frac{r}{530\ \mathrm{kpc}}\right)^2\right]^{-1}$$

and

2) an x-ray emitting, hot gas component of radial profile

$$\rho_x \sim \left[1 + \left(\frac{r}{380\ \mathrm{kpc}}\right)^2\right]^{-1}.$$

Their preferred resulting gravitational mass distribution

$$\rho_b \sim \left[1 + \left(\frac{r}{298\ \mathrm{kpc}}\right)^2\right]^{-1},$$

implies the presence of a dark matter component.

It has been pointed out by Sciama, Persic and Salucci (1992, 1993) that the above radial distributions imply that the dark matter is more concentrated than the gas and the galaxies. In order to demonstrate this they computed from the mass model a quantity γ, defined as the ratio between the mass in diffuse dark matter and the mass in gas and in galaxies (inclusive of their dark haloes) at two different radii. They chose as the outer radius 2 Mpc, where according to the model the total mass (for $h = 1/2$) is 10^{15} $M_\odot$, the gas mass is 2.8×10^{14} $M_\odot$ and the mass associated with galaxies (assuming with Hughes and Tanaka that $M/L_V = 5$) is 4×10^{13} $M_\odot$. The inner radius was chosen as 0.2 Mpc. They found that $\gamma \sim 2$ at 2 Mpc and $\gamma \sim 11$ at 0.2 Mpc. However, they considered that a more realistic value of M/L_V would be 20. They then found that γ goes from ~ 1 to ~ 10 over the same radial interval. These results show that in A665 the diffuse dark matter is more concentrated than the visible matter by an order of magnitude. Fig. 2.1 shows in detail the dark to visible density ratio as a function of radius. One can clearly see the central concentration of dark matter and that, for high values of the mass to light ratio of individual galaxies, the outer regions of the cluster are dominated by matter associated with visible structures. In this latter case the diffuse dark matter is not only more concentrated but resides almost entirely in the innermost regions of the cluster.

It should be emphasised that the relative distribution of dark and luminous matter in clusters is opposite to what is found in galaxies. In the well-studied case of NGC 3198 (van Albada *et al.* 1985), for example, the dark to luminous mass ratio increases by a factor ~ 30 when going from one half to five times the radius of the stellar disk. To highlight this dramatic difference we show in Fig. 2.2 (from Sciama, Persic and Salucci 1993) the local dark to visible density ratio plotted between one half and five times the typical size of the visible matter distribution for A665 and NGC 3198, using the mass model solutions of Hughes and Tanaka and of van Albada *et al.* respectively. If the differences highlighted by Fig. 2.2 hold generally, they may provide a clue to the identity of the dark matter in clusters and in individual galaxies. This question will be discussed in chapter 4.

2.3 Gravitational Lens Studies of Clusters

A totally different but even more model-independent probe of the dark matter in the central regions of clusters comes from Ein-

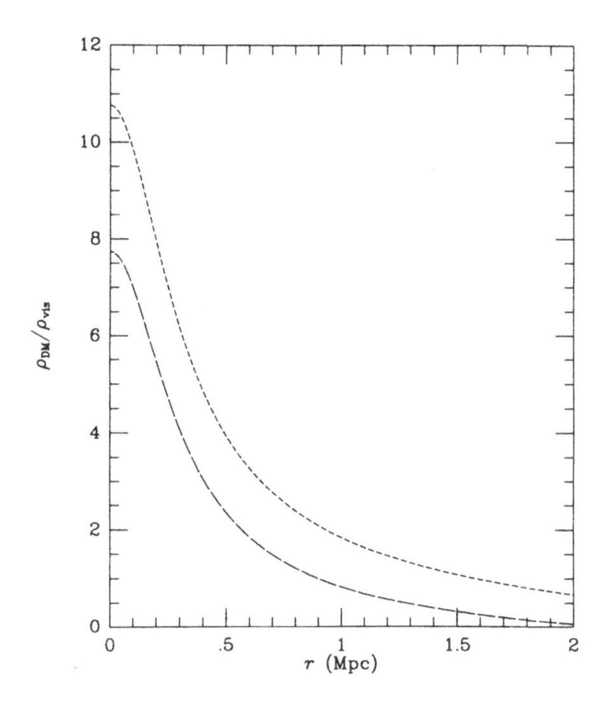

Fig. 2.1 The dark to visible density ratio in the galaxy cluster A665 as a function of radius. This plot is based on Hughes and Tanaka's (1992) mass model. It shows that the diffuse dark matter in A665 is considerably more concentrated than the distribution of hot gas and member galaxies. (The short-dashed line refers to an assumed value of the dynamical galaxy mass-to-light ratio of $M/L_v = 5$; the long-dashed line refers to $M/L_v = 20$). [From Sciama, Persic and Salucci 1992, 1993].

stein's gravitational lens effect. The idea here is that a rich cluster of galaxies can act as a gravitational lens for the optical radiation emitted by a galaxy lying behind the cluster on a caustic of the lens. A general discussion of this process has been given by Blandford (1990). The basic observational discovery was made in 1986 when a long circular arc with angular radius $\sim 25''$ and angular length $\sim 20''$ was found in the Abell cluster A370 by Lynds and Petrosian (1986) and by Soucail *et al.* (1987). The spectrum of this arc has a measured red shift of 0.72, about twice the cluster redshift of 0.37, and is believed to correspond to a highly distorted

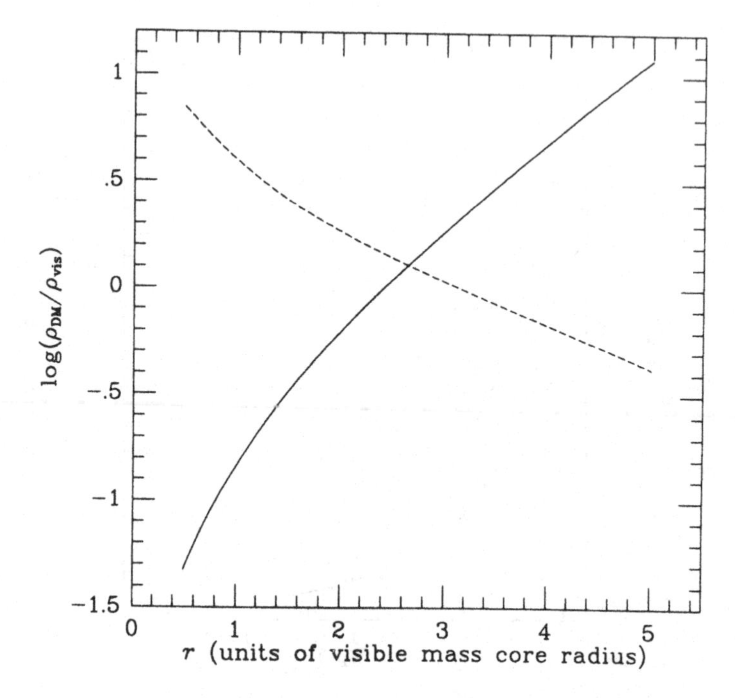

Fig. 2.2 The local (diffuse dark)-to-visible ratio as a function of radius (in units of the scale-size λ of the visible mass distribution) between $\frac{1}{2}\lambda$ and 5λ for the spiral galaxy NGC 3198 (solid line; van Albada *et al.* 1985) and for the galaxy cluster A665 (Hughes and Tanaka 1992). [From Sciama, Persic and Salucci 1993].

background galaxy.

The theory of this lensing effect has now been widely discussed in the literature e.g. Blandford (1990), Grossman and Narayan (1989), Hammer and Rigaut (1989), Bergmann *et al.* (1990), Blandford and Narayan (1992), Schneider *et al.* (1992), Miralda-Escude (1993). The essential point is that the location Θ_a of the arc is determined by the requirement that the mean surface density of the lensing cluster within Θ_a should be just equal to the critical density Σ_{crit}, which is given by

$$\Sigma_{crit} = \frac{c^2}{4\pi G}\frac{D_{OS}}{D_{OL}D_{LS}},$$

where O is the observer, L is the lensing cluster, S is the lensed source and D refers to angular diameter distances.

To see what is involved in this formula for Σ_{crit}, we may in the present case assume that each of D_{OS}, D_{OL} and D_{LS} is roughly equal to the radius of the universe c/H_0 (to within a factor of 2 or 3). Thus

$$\Sigma_{crit} \sim \frac{cH_0}{4\pi G}$$

$$\sim 0.1 \text{ gm cm}^{-2} \quad \text{or} \quad 500 \text{ M}_\odot \text{ pc}^{-2}.$$

Thus the observation of the arc in A370 is telling us directly that the mean column density in this cluster within 25″ of its centre is roughly 0.1 gm cm^{-2}. A more accurate value, based on actual distances, would be several times larger, and is also much larger than the column density of the visible matter. Thus we learn again that the gravitational field of the cluster is dominated by dark matter.

Grossmann and Narayan, and Bergmann *et al.* have also emphasised that for A370 the dark matter is distributed differently from the visible matter, and is again more concentrated. This emerges from their general modelling, but Bergmann *et al.* also give a simple argument for it which goes as follows. If the core radius of the dark matter were much greater than Θ_a, then the column density within this angular distance would be essentially constant, and it would be a remarkable coincidence if it were just equal to Σ_{crit}, which is determined by unrelated quantities (c, G and the distances D_{OS}, D_{OL} and D_{LS}). This probability argument has been strengthened recently by the discovery (Guhathakurta 1991) of several other clusters containing arcs.

On the other hand, if the core radius Θ_D of the dark matter is much less than Θ_a then, for an arc to appear, it is necessary only that the column density of the cluster should exceed Σ_{crit} at its centre. In this case the column density decreases with angular distance from the centre, and there must exist a position where the mean column density has decreased to Σ_{crit}. The arc would then appear at this position. Since the core radius of the visible matter in A370 is observed to exceed Θ_a ($\sim 25''$) it follows that the dark matter in this cluster is more concentrated than the visible.

A further point made by Bergmann *et al.* is that if $\Theta_D > \Theta_a$ the total mass of the cluster would be improbably large.

These results have been confirmed by Guhathakurta (1992) who studied 26 clusters and found that all the optically rich and/or compact ones show appreciable lensing. He also found that the dark matter core sizes derived from lensing for these objects are usually smaller than the x-ray core radii.

Further confirmation follows from the more model-dependent analysis of the widths of the arcs by Hammer (1991). The width of an arc in relation to the source size is determined by the value σ_a of the column density σ at the position of the arc. The most plausible assumptions about the source sizes imply that $\sigma_a \leq \frac{1}{2}\Sigma_{crit}$. This would again mean that $\Theta_D < \Theta_a$.

Very recently a report has appeared in Science (Flam 1992) that Richard Ellis of Durham University has observed a spectacular double image of a galaxy refracted by a foreground cluster of galaxies AC 114. According to this report Ellis has calculated that the cluster contains 50 times more dark matter than visible matter and that the dark matter clumps around the centre of the cluster. Ellis is quoted as arguing that the detail in the new image now makes the evidence for the clumping of dark matter much stronger.

We conclude from this discussion that in many clusters of galaxies the dark matter is more concentrated than the visible matter. This result will be important for us when we come to consider in chapter 4 the identity of the dark matter.

3
Dark Matter in Intergalactic Space

3.1 Introduction

The idea that there may be significant quantities of dark matter distributed smoothly throughout the universe as a whole developed gradually. In its modern formulation the idea is based on a number of considerations. The first concerns the role played by the mean density of the universe in the cosmological models of general relativity. These models have a fundamental status in discussions of cosmological dark matter, and so we devote much of this chapter to an account of them.

The second consideration concerns the mean density of ordinary matter in the universe. By "ordinary matter" I mean atoms, neutral or ionised, which are collectively referred to as baryonic. Estimates have been made of the contribution of visible baryons to the mean density of the universe using direct astronomical measurements. An estimate has also been made, using indirect arguments, of the total contribution of baryons, visible and invisible, to the mean density. This estimate is based on a comparison of the measured abundances of certain light elements (D, He^3, He^4 and Li^7) with the calculated output of thermonuclear reactions occurring in the "first three minutes" after the hot big bang origin of the universe. These arguments will also be described in this chapter. We shall find that, according to modern estimates, the mean density in visible baryons is significantly less than the total mean density in baryons. If these estimates are correct, an appreciable number of baryons must be dark.

The third consideration concerns the contribution of non-baryonic matter to the mean density of the universe. Various forms of more or less "exotic" matter have been proposed under this heading. None of these forms has yet been definitely observed by astronomers, so all of them would have to be dark. The general idea is that non-baryonic matter would have been created in the high density, high temperature regime of the early universe, and that some of it would have survived to the present day. The possibilities here include neutrinos of non-zero rest mass, and various objects predicted by speculative theories of elementary particle physics, such as photinos, axions and topological defects. It should be stressed that the very existence of all these objects (including neutrinos of non-zero rest mass) is quite speculative at the moment. Nevertheless we shall see that modern measurements of the total mean density of the universe significantly exceed the estimated density in baryons, visible and invisible. If correct, this would mean that some form of non-baryonic dark matter is essential. It is this conclusion which gives the cosmological dark matter problem its particular importance.

We begin our discussion with an extended account of Newtonian and relativistic models of the universe and describe the role of the mean density of the universe in these models. We then discuss the cosmic microwave background, the hot big bang and the primordial nucleosynthesis of D, He^3, He^4 and Li^7. Finally we consider modern estimates of the density, age and expansion rate of the universe.

3.2 Models of the Universe

3.2.1 The Newtonian Dynamics of a Large Gas Cloud

To prepare the way towards understanding relativistic cosmology we shall consider the Newtonian dynamics of a large gas cloud. Not only is the Newtonian theory mathematically simpler, it also leads to many results that are essentially the same as in relativity, as was discovered in 1934 by Milne and McCrea. The relation between the gas cloud on the one hand and the galaxies that make up the Universe on the other need not be specified too closely. One can think of the galaxies either as the particles of the gas, or as being localised condensations in an actual intergalactic (atomic or ionised) gas, condensations that act as tracers for the average motion and perhaps the average density of the gas in their general vicinity.

What is important is that the gas cloud should not be taken

to be infinitely large. As Newton discovered, his dynamics and gravitational theory run into difficulties when applied to an infinite system. For instance the gravitational potential at a point due to all the matter in the system would be infinite. This difficulty does not arise in general relativity, but here we may avoid it by taking our cloud to be large but finite in size. Another way of avoiding it was proposed in the last century by Neumann and Seeliger, who added to the Newtonian gravitational force a repulsive force directly proportional to the distance of a particle from the origin and independent of the physical properties of matter. We shall not adopt this device here.

There is one important difference between a large cloud and an infinite one. The large cloud has a unique centre while the infinite one does not. We do not really want a special point to be picked out in this way, and we can minimise the effect of having one by making the cloud uniform out to its edge, isotropic about its centre and much larger than any distance that has yet been measured. Under these circumstances any galaxy that we can detect would see around itself with arbitrarily high precision a uniform isotropic Universe. In the last analysis we can never distinguish observationally between an infinite Universe and a finite one that is suitably larger than any distance yet surveyed. The debatable assumption that we have made is then not so much that the system is finite as that it is uniform and isotropic. For the moment we may regard these assumptions as being useful for a first attack on the theoretical problem.

The assumptions of uniformity and isotropy are useful not only for permitting any point in the observable region of the cloud to regard itself as the centre. They also very much simplify the motion of the cloud as seen by any observer moving with the cloud. In fact the velocity v of a particle at a radius vector r from a co-moving observer then satisfies the simple relation

$$v = f(t)r, \tag{1}$$

where f is any function of the time t. Thus *at a given time* the motion of the particles with respect to any one co-moving particle satisfies a linear velocity-distance relation, a relation which is called the Hubble law. This result shows how stringent are the assumptions of homogeneity and isotropy. Of course we can regard this linear law as a first-order approximation to a more complicated law, but it does seem to be a reasonable approximation to what is observed. In fact when we test it out to distances large enough for a term in r^2 to perhaps become appreciable we must also allow for the fact that we may be looking so far back in time that the quantity $f(t)$ can no longer be taken to be constant. Thus we could absorb a small non-linear term by slightly modifying f.

It is useful for our further development to integrate the velocity law (1) to give the position of the particle at time t. The result is

$$r = R(t) r_0, \tag{2}$$

where $R(t)$ is related to $f(t)$ by the equation

$$\frac{1}{R}\frac{dR}{dt} = f(t), \tag{3}$$

as is easily seen by differentiating (2) and comparing it with (1). In (2) r_0 is the position of the particle at some standard time t_0, and so

$$R(t_0) = 1.$$

We see from (2) that the only possible motions consistent with uniformity and isotropy are those of uniform expansion or contraction, a simple scaling up or down with a time-dependent scale factor $R(t)$. To simplify the notation we shall write (3) in the form

$$\frac{\dot{R}}{R} = H(t)$$

so that

$$v = Hr. \tag{4}$$

Notice that the Hubble constant H is independent of r (that is what is meant by the Hubble law) but does depend on t.

So far we have managed without using Newton's second law of motion or his law of gravitation. Everything has followed from our symmetry assumptions alone, and we may regard our results up to this point as kinematical in character. They show that the whole motion of the cloud is determined by just one arbitrary function of the time. To determine this function more closely we must introduce dynamical considerations. This problem is much simplified by the well-known Newtonian result that in a uniform isotropic system the gravitational force acting on a particle at the position r relative to the centre is entirely due to the matter lying closer to the centre than does the particle. We can use this fact to obtain a simple result for the dynamics of the cloud as seen from its centre, and we can transfer this result to any other origin within the observable region by using the relation (4). In this way we avoid the tricky question of whether all the co-moving observers, who will be in acceleration relative to one another, represent inertial frames of reference.

Newton's laws lead to the following equation for the scale-factor $R(t)$:

$$R^2\ddot{R} + \frac{4\pi}{3}G\rho(t_0) = 0, \tag{5}$$

where G is the Newtonian constant of gravitation and $\rho(t_0)$ is the density of the cloud at the standard time t_0 which, because of the conservation of matter, satisfies the equation

$$\rho(t) = \frac{\rho(t_0)}{R^3(t)}, \tag{6}$$

since $R(t_0) = 1$. Our dynamical equation (5) for R shows immediately what is intuitively obvious: that we cannot have a static cloud ($\dot{R} = \ddot{R} = 0$) unless $\rho = 0$. We are thus led to expect a systematic motion of expansion or contraction on a scale over which the Universe is approximately uniform. In fact this result, translated into relativistic terms, was derived rigorously from general relativity before the Hubble law was established observationally. It might seem at first sight that a star or planet, which are at least quasi-static, constitute counter-examples. However, in these cases gravity is balanced by a pressure gradient, which cannot exist in a uniform system. In any case the enormous pressure gradient needed to stabilise the Universe is obviously not in fact present.†

Fortunately the equation for $\ddot{R}$ can be integrated easily to give a dynamical equation for $\dot{R}$, the rate of expansion or contraction. To effect this integration we multiply (5) by $\dot{R}/R^2$ and so obtain the integrated equation

$$\dot{R}^2 = \frac{8\pi}{3}\frac{G\rho(t_0)}{R} - k. \tag{7}$$

† Moreover in relativity the pressure contributes (positively) to the gravitational field. In an ordinary star this contribution is negligible, but a pressure gradient which at first sight might stabilise the Universe would actually increase the effective gravity.

Here k is a constant of integration which is a measure of the total energy (kinetic plus potential) of a particle. With the sign we have chosen for k (which is the conventional one) the cloud is gravitationally bound or unbound according as k is positive or negative. When k is zero the kinetic and potential energies are equal and opposite, and the cloud can just expand to infinity. We now consider these three cases in more detail. This is worth doing *because relativity leads to the same equation for the scale-factor* $R(t)$, although the constant k then has a somewhat different meaning.

(i) $k = 0$. In this case

$$\dot{R}^2 = \frac{8\pi}{3}\frac{G\rho(t_0)}{R},$$

and $\dot{R}$ tends to zero as R tends to infinity. This equation for $\dot{R}$ can be integrated to give the explicit time-dependence of R. We find that

$$R \propto t^{\frac{2}{3}},$$

where the constant of proportionality is $6\pi G\rho(t_0)^{\frac{1}{2}}$. The graph of $R(t)$ is shown in Fig. 3.1. In relativity this relation characterises the well-known Einstein-de Sitter model of the Universe.

(ii) $k > 0$. In this case the cloud is gravitationally bound and expands out to a maximum size R_{max} which occurs when $\dot{R} = 0$, that is, when

$$R_{max} = \frac{8\pi}{3}\frac{G\rho(t_0)}{k},$$

at which point the motion reverses into a collapse. This collapsing phase of the motion is familiar to astronomers concerned with star formation, since they are interested in a finite cloud collapsing from rest. The solution of (7) is fairly simple and one finds that the $R(t)$ curve is a cycloid, as in Fig. 3.2. This gives us an oscillating Universe, although there is no justification in adding further cycles to the oscillation as some writers do.

(iii) $k < 0$. In this case

$$\dot{R}^2 = \frac{8\pi}{3}\frac{G\rho(t_0)}{R} + (-k) \quad (-k \text{ a positive constant}). \qquad (8)$$

Hence as R tends to infinity the rate of expansion $\dot{R}$ tends to a non-zero positive quantity. In other words, the particles have

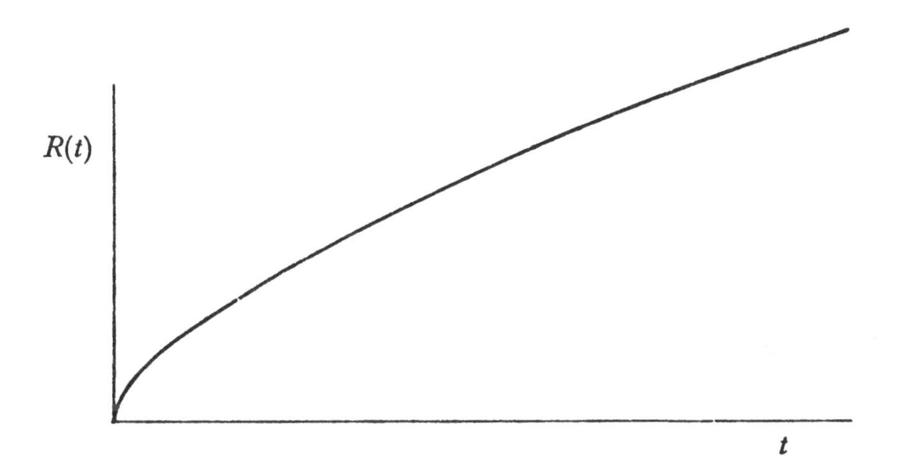

Fig. 3.1 The scale factor of the Universe $R(t)$ for a Newtonian model with zero total energy, or a relativistic model with zero pressure and space curvature (the Einstein-de Sitter model).

excess kinetic energy and are still moving apart when the cloud is infinitely large and dilute. Unfortunately (8) cannot be integrated in simple terms, but R does depend simply on t for both small and large t. For small t, R is small and the first term on the right hand side of (8) dominates over the second term. This gives us the same equation as in case (i) with $k = 0$, and we have

$$R(t) \propto t^{\frac{2}{3}} \quad (t \text{ small}).$$

For large t, R is large and the second term dominates. The equation then integrates to give

$$R \propto t \quad (t \text{ large}),$$

which corresponds to unaccelerated expansion, gravitation being negligible, for large t. A sketch of $R(t)$ over the whole range of t is given in Fig. 3.3.

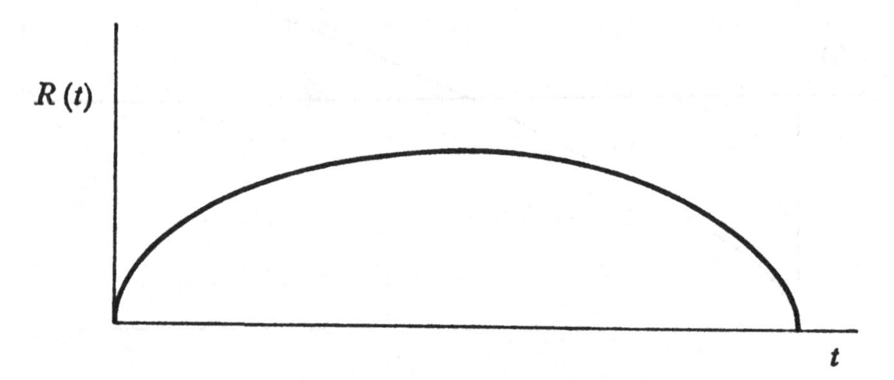

Fig. 3.2 The scale factor of the Universe for a gravitationally bound model.

We now consider some of the observable properties of these models in order to see which fits the actual Universe best.

(*a*) *Expansion rate.* We have already seen in (4) that the Hubble constant H is given by

$$H = \dot{R}/R.$$

Its present value is still very controversial, and this controversy is discussed later. The range of possible values is usually given as

$$H_0 = (10^{10}\text{ year})^{-1} h = 100\ h\text{ km sec}^{-1}\text{ Mpc}^{-1}$$

where

$$0.5 \leq h \leq 1.$$

If we choose the present time as the standard time t_0, then the present value of R is unity, and the present value of H is $\dot{R}$. We thus know that the slope of the $R(t)$ curve at present is $(10^{10}\text{ years})^{-1}\ h$.

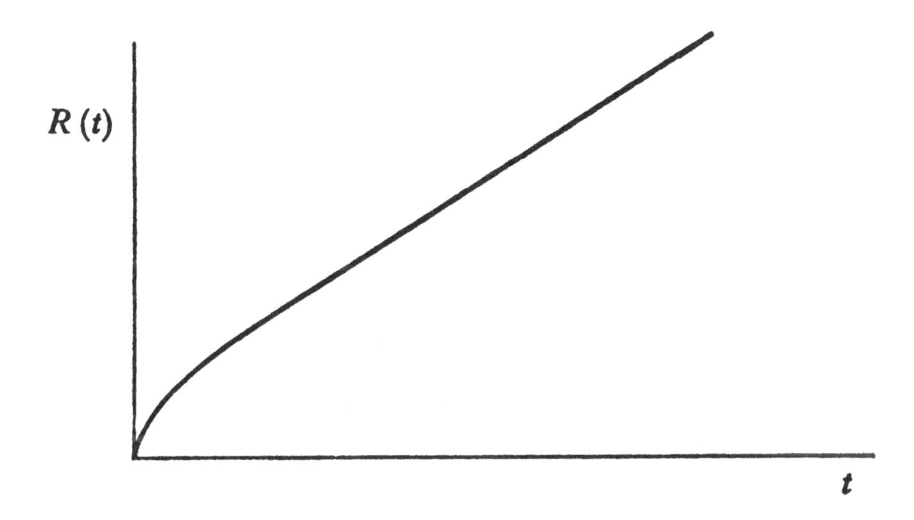

Fig. 3.3 The scale factor of the Universe for a gravitationally unbound model.

(*b*) *The present age of the Universe.* Fig. 3.4 shows geometrically that the time in the past when R was zero, which we may call the age of the Universe, is less than H_0^{-1} in all the models. This is obvious physically since the Universe was expanding faster in the past because of its self-gravitation (see (c) below). In the Einstein-de Sitter model ($k = 0$) we have $R \propto t^{\frac{2}{3}}$, and so

$$t_0 = \frac{2}{3}\frac{R}{\dot{R}} = \frac{2}{3H_0}.$$

If $H_0 \sim (10^{10}\ \text{years})^{-1}$, we would have $t_0 \sim 6.7 \times 10^9$ years, which is somewhat less than the ages of the oldest stars in the Galaxy, and dangerously close to the age of the Sun. This time-scale difficulty would be relieved if the actual value of the Hubble constant were somewhat less than (10^{10} years^{-1})(corresponding to h close to 0.5).

The time-scale difficulty is even worse in the oscillating models

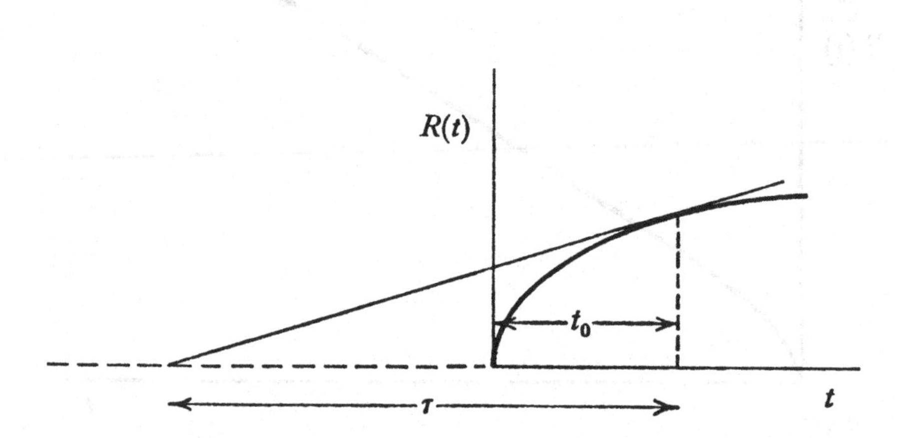

Fig. 3.4 The age of the Universe t_0 is always less than the reciprocal of the Hubble constant H_0 (which is determined by the tangent to the $R(t)$ curve).

($k > 0$), where $t_0 < 2/(3H_0)$. By contrast in the ever-expanding models ($k < 0$), t_0 approaches $1/H_0$ more nearly as the present density $\rho(t_0)$ is lowered, the effects of gravity are correspondingly reduced, and the approximation $R \propto t$ is valid from an earlier starting time.

(*c*) *Deceleration parameter.* In all our models the expansion of the Universe is decelerating because of self-gravitation. The amount of deceleration gives us a useful measure of the amount of self-gravitation and so of the material density. The deceleration is essentially $-\ddot{R}$, but it is convenient to define it in such a way that it is independent of the time t at which we set $R(t_0) = 1$, and also such that it is dimensionless. We achieve the first aim by considering the quantity $-\ddot{R}/R$. This has dimensions t^{-2}, so to obtain a dimensionless quantity we multiply by $R^2/\dot{R}^2 (= H^{-2})$.

We thus take as our definition of the deceleration parameter q the equation

$$q = -\frac{R\ddot{R}}{\dot{R}^2}.$$

This is a purely kinematical definition. If we adopt the dynamical laws (5) and (6) and use (4) we find

$$q = \frac{4\pi}{3}\frac{G\rho}{H^2}, \tag{9}$$

which shows the way in which q measures the density of matter ρ. In general q changes with time but an interesting exception is the Einstein-de Sitter model ($k = 0$) in which $R \propto t^{\frac{2}{3}}$, and so $q = \frac{1}{2}$ at all times. We have in fact the following relations

$$\begin{array}{llll} \textit{model} & (i) & (k = 0) & q = 1/2 \\ \textit{model} & (ii) & (k > 0) & q > 1/2 \\ \textit{model} & (iii) & (k < 0) & q < 1/2 \end{array}$$

(*d*) *Density.* This is closely related to the deceleration parameter, as we have just seen.

Model (i) ($k = 0$). We have $q = \frac{1}{2}$ and so from (9)

$$\frac{8\pi}{3}G\rho = H^2.$$

Thus this model has the important property of leading to a definite value for the present density once the present value of the Hubble constant H is known. This special value of the density is known as the critical density ρ_{crit}, where

$$\rho_{crit} = 2 \times 10^{-29}\, h^2 \text{ gm cm}^{-3}.$$

Other values for the density are then usually parametrised by Ω, where

$$\Omega = \rho/\rho_{crit}.$$

We shall see later that if $\Omega = 1$ there would be a cosmological dark matter problem.

Model (ii) ($k > 0$). In this type of model gravity is dominant and the density exceeds the critical value ($\Omega > 1$), thus accentuating the dark matter problem. It also suffers most from the time-scale difficulty, as we have seen.

Model (iii) ($k < 0$). In this type of model gravity is more or less unimportant except in the earliest stages and the present density can have any value less than the critical one ($\Omega < 1$). Of course observations give a lower limit for the density, which as we shall see corresponds to $\Omega \sim 0.1$. This lower limit is small compared with the value expected in the Einstein-de Sitter model. If it were close to the actual value the Universe would now be expanding at a rate which would remain nearly constant in the future. Such a model has the advantage that its present age is close to H_0^{-1}, and so is the nearest to solving the time-scale difficulty if $h \sim 1$.

This completes our discussion of the Newtonian dynamics of a large gas cloud. The reader will have noticed one very important omission. We have not discussed the behaviour of light in these models, and have therefore said nothing about the red shift, the apparent brightness of a distant source, and so on — the very properties that most closely link together observation and theory. The reason for this omission is that Newtonian theory does not provide us with a satisfactory account of the behaviour of light. Although the relativistic formula for the red shift is a very simple one, attempts to derive it in a Newtonian setting give one the uneasy feeling that one is stretching the Newtonian concepts too far. As a result one cannot rely on the answer until it has been checked by relativity. Under these circumstances there seems to be no point in studying further our over-simplified picture. We must therefore now turn our attention to relativistic cosmology. The discussion necessarily becomes somewhat more mathematical, and if the reader is uninterested in following the details he is advised to turn to page 35 which gives a summary of the results we describe.

3.2.2 Relativistic Cosmology

General relativity differs from the Newtonian theory of gravitation in the following respects:

(i) It is based on ten potentials instead of one (or Maxwell's four in electrodynamics).

(ii) It is a non-linear theory, the total effect of several bodies not being the simple sum of their separate effects.

(iii) Pressure as well as density is a source of gravitation.

(iv) It is usually expressed in geometrical language, the ten potentials giving the metrical properties of space-time, which is curved in the presence of a gravitational field.

(v) There is no difficulty in having a gas cloud fill the whole of space.

Some of these differences can lead to great mathematical difficulties, but fortunately these are minimised when we impose the powerful symmetry assumptions of uniformity and isotropy as we did in the Newtonian discussion. These symmetry assumptions alone limit the metric (which gives the four-dimensional distance between neighbouring points of space-time) to the following form:

$$ds^2 = c^2dt^2 - \frac{R^2(t)}{(1+\frac{1}{4}kr^2)^2}\left[dr^2 + r^2(d\theta^2 + \sin^2\theta d\phi^2)\right],$$

as was shown by Robertson and Walker following the pioneering work of Milne. This differs from the special relativity metric for Minkowski space-time only by the presence of the undetermined scale-factor $R(t)$ and the constant k. It is clear in a general way that the scale-factor $R(t)$ has much the same meaning here as in the Newtonian theory. To see this consider the Universe at a particular time t_0. Then we have $dt = 0$, and the metric for 3-dimensional space at the time t_0 becomes

$$ds^2 = -\frac{R^2(t_0)}{(1+\frac{1}{4}kr^2)^2}\left[dr^2 + r^2(d\theta^2 + \sin^2\theta d\phi^2)\right].$$

At a later time t_1 we would have exactly the same metric, except that every interval ds would be multiplied by the factor $R(t_1)/R(t_0)$. If this factor is greater than 1, intervals are increasing with time and we have an expanding Universe.

The meaning of the quantity k is rather different here, however. Gravitational potential energy is an elusive concept in general relativity, and it is best to think of k as giving the *curvature* of 3-dimensional space at any time t_0. We then have the following possibilities:

(i) $k = 0$. Three-dimensional space is then Euclidian. In particular the surface area of a sphere of radius r is $4\pi r^2$.

(ii) $k > 0$. In this case the geometry of space is said to be spherical. It is in fact the 3-dimensional analogue of the geometry on the surface of a sphere. On such a surface a circle is the locus of points at a constant distance from a given point, distance being measured along great circles (Fig. 3.5). The circumference of such a circle is less than 2π times its radius. The difference

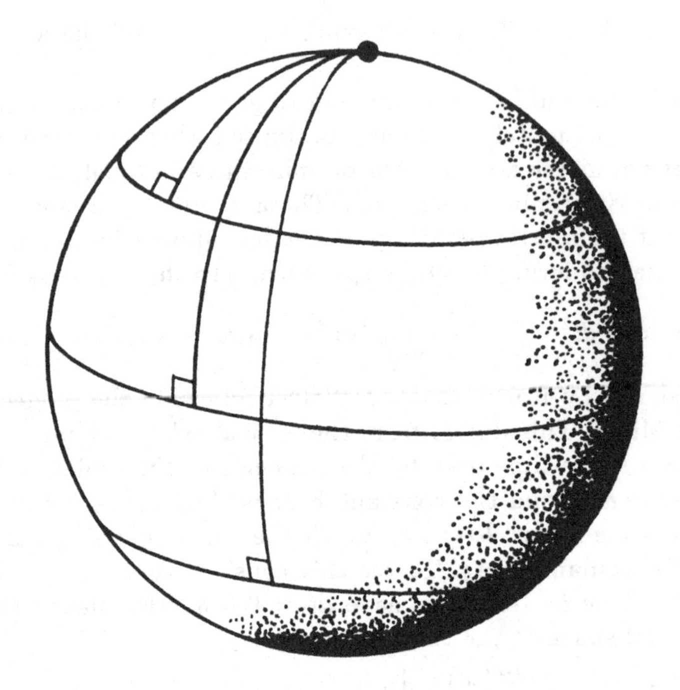

Fig. 3.5 The circumference of a circle on a sphere is less than 2π times its radius (as measured along part of the great circle). The corresponding geometry is non-Euclidean, and has positive curvature.

is small when the radius is small, but for a larger radius it becomes substantial until the radius goes one-quarter way around the sphere, when the circumference is a maximum. For a larger radius still the circumference gets smaller again, and goes to zero as the radius goes half way round the sphere. In a similar way in the 3-dimensional spherical space the surface area of a sphere of radius r is less than $4\pi r^2$, grows to a maximum as r increases and then shrinks to zero again. The volume of the space is finite and is in fact equal to $\pi^2 R^3$, so it increases as R increases with time.

(iii) $k < 0$. In this case the geometry of space is said to be hyperbolic. The surface area of a sphere of radius r is greater than $4\pi r^2$. The volume of the space is infinite except in pathological cases that need not concern us.

Our results so far have been entirely kinematical. They place no restriction on R as a function of time, nor do they impose any relation between $R(t)$ and k. To proceed further we must use the relativistic analogue of Newton's law of gravitation, that is, we must use Einstein's field equations. At one time, before the expansion of the Universe had been discovered, Einstein proposed a modification of his field equations that would permit the Universe to be static (R = constant). The extra term involved an undetermined constant (the cosmological constant) and for a suitable choice of its sign its effect would be to oppose self-gravitation and so permit a static solution (as in the Neumann-Seeliger modification of Newtonian theory). The value of the constant required to achieve this situation is so small that its presence in the field equations would not disturb the agreement between general relativity and observation in the solar system. In order to keep things simple, we shall not introduce a cosmological constant into the field equations. A modern discussion of this constant and its relation to quantum field theory has been given by Weinberg (1989).

In view of the fact that in relativity pressure acts as a source of gravitation we must be careful now to specify the pressure in the gas cloud that represents the matter in the Universe. This pressure can be taken to include contributions from the peculiar motions of the galaxies, from the intergalactic gas (which may be hot), from radiation, and from intergalactic magnetic fields and cosmic rays. At the present time these pressures are almost certainly unimportant as a source of gravitation in comparison with the energy-density of the matter in galaxies. However, for the moment we shall keep the pressure p in as an unknown. Einstein's field equations now lead to the following relations:

$$\dot{R}^2 = \frac{8\pi}{3} G\rho R^2 - k, \tag{10}$$

$$2\frac{\ddot{R}}{R} + \frac{\dot{R}^2}{R^2} = -8\pi\frac{Gp}{c^2} - \frac{k}{R^2}. \tag{11}$$

The first of these equations resembles our Newtonian equation (7) and would indeed be identical to it if we could write, as before,

$$\rho(t) = \rho(t_0)/R^3.$$

However, we must remember that if the pressure does work during the expansion, this work will change the energy density and so, in relativity, it will change the matter density ρ. At this point then we set $p = 0$ for simplicity (it will be important later to reintroduce it). If we now multiply (11) by $R^2\dot{R}$ we get

$$2R\dot{R}\ddot{R} + \dot{R}^3 = -k\dot{R},$$

which integrates immediately to

$$R\dot{R}^2 = -kR + \text{constant}.$$

Comparison with (10) now shows that

$$\rho R^2 = \frac{\text{constant}}{R},$$

so that $\rho \propto 1/R^3$, as we want. Thus in the case of zero pressure the governing equation for the scale-factor R is

$$\dot{R}^2 = \frac{C}{R} - k,$$

where

$$C = \frac{8\pi}{3} G\rho R^3 = \text{constant}.$$

Hence despite all the differences between general relativity and Newtonian theory, *the scale-factor R satisfies the same equation in both theories so long as the pressure is negligible.* This is the Milne-McCrea theorem. Accordingly, the classification of models and the time-dependence of R is the same in both theories, and we need not repeat our discussion of these questions. Since in the present case the gas cloud fills the whole Universe at all times, it is better not to speak of bound and unbound clouds for $k > 0$ and $k < 0$, but rather of spherical and hyperbolic space, or closed and open space, or oscillating and ever-expanding space.

Two further models that are not pressure-free deserve mention at this point. One concerns the important physical case in which radiation dominates completely over matter as a source of gravitation. This is the situation that existed in the early stages of our own Universe, as we shall see. The pressure was no longer negligible in those stages and in fact $p/c^2 = \frac{1}{3}\rho$. We can eliminate ρ by adding (10) and (11) to obtain

$$\frac{\ddot{R}}{R} + \frac{\dot{R}^2}{R^2} + \frac{k}{R^2} = 0.$$

As before we can neglect k for sufficiently small R, and then we can integrate the equation to obtain

$$R \propto t^{\frac{1}{2}} \quad (t \text{ small}).$$

This corresponds to a more rapid expansion than when only pressure-free matter is present ($R \propto t^{\frac{2}{3}}(t$ small)), because the pressure of radiation exerts its own gravitational field, thereby increasing the amount of gravity acting. This increases the rate of expansion, as is obvious if we reverse the sense of time and consider the resulting rate of collapse.

The second model with pressure which we wish to mention has $p/c^2 = -\rho$, that is, it contains a tension rather than a pressure. The corresponding gravitational effect is now repulsive, and in fact we obtain a model whose expansion is accelerating rather than decelerating. We shall in addition take $k = 0$ to obtain the model we want. We can then eliminate ρ from (10) and (11) to obtain

$$\frac{\ddot{R}}{R} - \frac{\dot{R}^2}{R^2} = 0,$$

or

$$(\ln \ddot{R}) = 0.$$

Hence

$$\ln R = t/\tau + b \quad (\tau, b \text{ constants}),$$

and

$$R \propto e^{t/\tau}.$$

This model differs from our previous ones in that R does not go to zero a finite time ago (Fig. 3.6). It is the celebrated de Sitter model (not to be confused with the Einstein-de Sitter model which has $R \propto t^{\frac{2}{3}}, k = 0$). The exponential curve is self-similar, that is, one cannot tell where one is along it by intrinsic measurements; it has no natural origin. That is why the de Sitter metric forms the basis of the steady state theory of Bondi and Gold (1948) and Hoyle (1948). It is also the basis of early forms of the inflationary theory (Guth 1981, Linde 1990, Narlikar and Padmanhaban 1991).

In the steady state theory the Universe does not evolve from a dense state to a dilute one. The reason is that (10) with $k = 0$

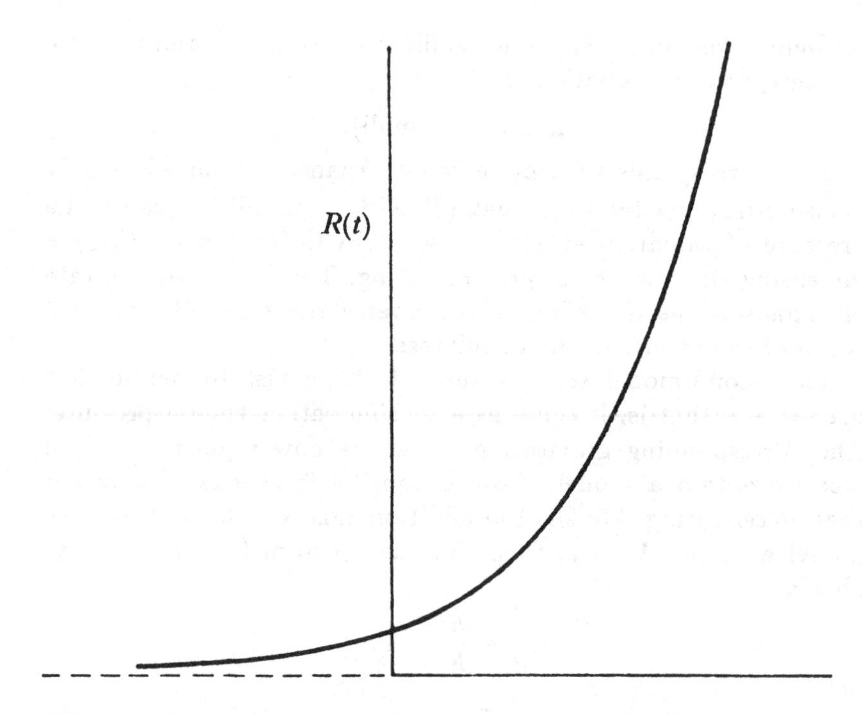

Fig. 3.6 The scale factor of the Universe in the de Sitter (steady state) model.

would give

$$\frac{8\pi}{3}G\rho = H^2.$$

Although this is the same relation as in the Einstein-de Sitter model, there is an important difference, for here H is independent of time and so ρ is also independent of time, whereas in the Einstein-de Sitter model $H \propto t^{-1}$ and $\rho \propto 1/t^2$. In the steady state model the density ρ remains constant because the work done by the tension during the expansion results in a continuous 'creation' of matter. This just compensates for the diluting effect of the expansion.

The steady state model is attractive in many ways, but has been discarded in favour of big bang models mainly (but not entirely) because of the discovery of the 3 K cosmic microwave background discussed below. However, some of its main ideas have been resuscitated in the inflationary theory, where they would prevail in the period $\sim 10^{-35}$ to 10^{-32} seconds after the big bang.

3.2.3 Summary of Newtonian and Relativistic Cosmology

Our discussion of Newtonian and relativistic cosmology has been rather elaborate, and some readers will probably be content with a brief summary that would suffice to make the rest of this book intelligible. We therefore provide such a summary here before going on to consider the propagation of light in the various models.

We assume the Universe to be uniform and isotropic. Its behaviour is then governed by one function of time $R(t)$ and one constant k. We call $R(t)$ the scale-factor of the Universe; it can be thought of as giving the time-dependence of the distance between two particles (galaxies) and therefore governs the rate of expansion of the universe. It does not matter which two particles we take because of the uniformity we are assuming. The quantity k in the Newtonian theory gives the total energy, kinetic plus potential, of a particle and so specifies whether the matter in the Universe forms a gravitationally bound system and so whether the expansion continues indefinitely. In relativity theory k determines the curvature of 3-dimensional space at any time, and also whether the expansion continues indefinitely.

There are four types of model universe in which we are mainly interested. The first three are pressure-free:

(i) $k = 0$ (Einstein-de Sitter model). In this case

$$R(t) \propto t^{\frac{2}{3}},$$

$$6\pi G\rho t^2 = 1,$$

$$t = \frac{2}{3}H^{-1},$$

and

$$\frac{8\pi}{3}G\rho = H^2,$$

where H is the Hubble constant $\dot{R}/R$. The present value of H^{-1} is $10^{10}\ h^{-1}$ years ($\frac{1}{2} \leq h \leq 1$) and so the present age since the moment of infinite density at $t = 0$ is $6.7 \times 10^9\ h^{-1}$ years, and the present density is $2 \times 10^{-29}\ h^2$ gm cm^{-3}. This is known as the critical density.

(ii) $k > 0$ (oscillating model). In this case $R(t)$ is a cycloid (Fig. 3.2), the present age is less than in the Einstein-de Sitter model, and the density is greater.

(iii) $k < 0$ (ever-expanding model). In this case $R(t)$ begins like $t^{\frac{2}{3}}$ and ends like t (Fig. 3.3); that is, the model ends up freely expanding with essentially no gravity holding it back. Its age now is greater than in the Einstein-de Sitter model and its density is less. In fact its age comes closer to H_0^{-1} the smaller is the present value of the density (which cannot, of course, be reduced below the density contributed by known galaxies).

The fourth model is of interest in the earliest stages of the expansion, when radiation may dominate completely over matter, and radiation pressure must be taken into account.

(iv) Radiation-filled model. In this case

$$R(t) \propto t^{\frac{1}{2}} \quad (t \text{ small}).$$

3.2.4 Red Shift

Non-technical books on cosmology usually give a simplified account of the red shift associated with the theoretical model universes. This procedure was reasonable at a time when observed red shifts were very small, but would be misleading now when red shifts as large as 5 have been observed, and when still larger ones may yet be discovered. The same remark applies to other observable properties of a distant source such as its apparent brightness (optical or radio) and its angular size. We are now in the regime where to analyse the observations we need rather accurate estimates of fully relativistic effects.

We shall not give the detailed calculations here but simply quote the results. Two points are worth noting, however. The first is that an object which partakes exactly of the motion of the substratum has constant r, θ, ϕ coordinates in the Robertson-Walker metric. These are thus co-moving co-ordinates of a type often adopted in fluid dynamics (where the phrase 'following the motion' is commonly used). The fact that such an object is receding from us is then contained in the function $R(t)$ which must, at the moment, be an increasing function of time. The second point is that distant objects are detected by the electromagnetic radiation (optical, radio, or x-) which they emit. The path of such radiation through space-time is given in relativity by a line of zero interval between any pair of points along it, that is, by putting $ds = 0$ in the Robertson-Walker metric. This enables us to calculate the time t at which radiation was emitted that we receive now (at time t_0) from a given source.

These considerations lead to the following simple formula for the red shift of a source which emitted its radiation at time t:

$$\frac{\lambda_0}{\lambda} = 1 + z = \frac{R(t_0)}{R(t)}.$$

The observed wavelength λ_0 and the emitted wavelength λ are in the same ratio as the scale-factors of the Universe at the moment of observation and the moment of emission. To relate this result to the Hubble law we consider a nearby source for which t differs little from t_0 ($t = t_0 - \delta t$). Then we can write

$$1 + z = \frac{R(t_0)}{R(t_0 - \delta t)} \sim 1 + \delta t \frac{\dot{R}(t_0)}{R(t_0)}.$$

Since for small z the classical Doppler formula is a good approximation, we have

$$z \sim \frac{v}{c}$$

and so

$$v \sim c \delta t H_0$$

from (4), which is just the Hubble velocity-distance law with distance given by $c\delta t$. By contrast, if z is large we cannot use this approximation. In special relativity z is related to v as follows:

$$1 + z = \left(\frac{c + v}{c - v} \right)^{\frac{1}{2}},$$

so a red shift of 5 would correspond to a velocity of 95 per cent of the speed of light. This relation between z and v is sometimes used in describing the significance of a large z, but since one should be using the Robertson-Walker metric this is really rather misleading.

If we know the red shift of an object we do not know very much about it if the function $R(t)$ for the actual Universe is unknown, which is in fact the case. All we can say is that when the radiation was emitted the Universe was more contracted than it is now by a factor $(1+z)$, so the density in the Universe was then greater than it is now by the factor $(1+z)^3$. For $z = 5$ this factor is 216, which is very substantial. If we knew the time at which the radiation was emitted we could combine this information with the red shift to derive the function $R(t)$, but we are unable to do this at the moment.

An alternative method is to determine the apparent optical or radio luminosity of the source which would give a measure of its distance and so of its light-time away. However, the red shift itself influences the apparent luminosity (a receding source being fainter than a stationary one of the same absolute luminosity). To deal with this situation it is convenient to introduce the concept of luminosity-distance.

3.2.5 Luminosity-Distance

If a source of absolute luminosity L has an apparent luminosity l then its luminosity-distance D is defined to be

$$D = \left(\frac{\mathrm{L}}{4\pi l}\right)^{\frac{1}{2}},$$

where L and l refer to the total radiation over all frequencies. In practice observations are made with a limited bandwidth and allowance must be made for the spectrum of the source because of the red shift involved. The definition of luminosity-distance is chosen so that the ordinary inverse square law is satisfied in terms of it. It can then be shown that

$$D = R(t_0)(1+z)\frac{r}{1+\frac{1}{4}kr^2}. \tag{12}$$

Since the co-ordinate r is not directly observable it is convenient to eliminate it using the equation for a light-ray. We then find that

$$D = \frac{c}{H_0 q_0^2}\left[q_0 z + (q_0 - 1)[(1+2q_0 z)^{\frac{1}{2}} - 1]\right]. \tag{13}$$

For small z this reduces to

$$D = \frac{zc}{H_0}\left[1 + \frac{1}{2}(1 - q_0)z\right].$$

The first term in z gives us back the Hubble law again, as we would expect since, for small z, luminosity–distance and light–time distance hardly differ. If the observations could be carried through to the higher terms in z we would have a means of determining the present value of the deceleration–parameter q_0 and so of the density ρ and the sign of the curvature k. It has not yet been possible to carry out this procedure in a reliable way, mainly because the evolutionary change of L with cosmic epoch has a dominating effect.

3.2.6 Angular Diameters

Another observational method of distinguishing between the different cosmological models is to study the dependence on red shift of the angular diameters of a class of objects that hopefully have a well-defined linear diameter. At very small distances the angular diameter would, of course, be inversely proportional to the distance, but for sources whose red shift is appreciable important relativistic effects come into play. To determine these effects we put $dr = dt = d\phi = 0$ in the Robertson-Walker metric and solve for $d\theta$:

$$d\theta = a\frac{(1 + \frac{1}{4}kr^2)}{rR(t)},$$

where we have written a for ds, the linear diameter of the source, and t is the time when the radiation was emitted. A neater expression for $d\theta$ is obtained if we introduce the luminosity-distance D from (12) and express $R(t)$ in terms of the red shift z. We then obtain

$$d\theta = \frac{a(1 + z)^2}{D}.$$

We can regard this as a relation between angular diameter, red shift and q_0 since the luminosity-distance D is known in terms of the red shift and q_0 from (13). In principle, then, we could determine q_0 from the observed $d\theta, z$ relation. In practice this is very difficult since the linear diameters may change systematically with z.

Despite this difficulty the $(d\theta, z)$ relation is of great interest. At large z ($z \gg 1$), D is nearly proportional to z and so $d\theta$ is nearly proportional to z. In other words at large z the angular diameter actually *increases* with increasing red shift and luminosity-distance. This is a consequence of the 'bending' of light in the curved space-time of the Robertson-Walker models, which thus have a lens-like behaviour. Of course, at small red shifts the angular diameter decreases as the red shift increases and so the angular diameter goes through a minimum. In the Einstein-de Sitter model, for example, this minimum occurs at a red shift of less than 2.

It must also be stressed that this remarkable behaviour of the angular diameters depends on the assumption that there is a substantial and essentially uniform distribution of matter between the galaxies. If, in fact, most of the matter in the Universe is contained within galaxies, and if the space along a line of sight between us and a distance source is essentially empty, then the angular diameter of the source would behave somewhat differently, and in particular would not go through a minimum.

Recently it has been claimed by Kellermann (1993) that measurements using very-long-baseline interferometry (VLBI) of compact radio sources associated with active galaxies and quasars may be largely free of evolutionary effects even at substantial red shifts. The relation which he obtained between angular size and red shift for a sample of these sources does indeed flatten off as expected for a smooth universe, and would imply that $q_0 \sim 0.5$ (see Fig. 3.8). This result still needs to be confirmed.

3.2.7 Number Counts and Radiation Backgrounds

If a class of objects such as galaxies or radio sources is distributed uniformly in space at each cosmic epoch, one can calculate from the Robertson-Walker metric the relative number of those objects that have a given red shift or a given measured brightness. We shall not give the relations here because attempts to compare them with observation have been frustrated by selection effects in the case of galaxies and evolutionary effects in the case of radio sources. These evolutionary effects arise because the fainter, more distant, objects are being seen at earlier epochs, when their intrinsic properties may have been (and probably were) different from their present properties. This effect can easily dominate over the differences between the different Robertson-Walker models. Moreover we do not understand the objects sufficiently well to know how they do in fact evolve. The observations we can make thus contain evolutionary effects and cosmological effects mixed up in unknown proportions.

A further aspect of the number counts that is of great practical importance concerns the contribution of all the sources to the unresolved background level of radiation. If one measures the flux of radiation reaching the earth in a solid angle much larger than that of an individual source, one is measuring this background. Sources that may be too faint to be detected individually can, in the aggregate, make a substantial contribution to the background. Thus to calculate the background it is necessary to know the relative number of sources of different absolute luminosity and to compute their contributions out to rather large red shifts.

This last point is often referred to as Olbers' Paradox. This paradox arose about 200 years ago from an attempt to compute the contribution of all the stars (or as we would now say, the galaxies) to the background light of the night sky. If we forgot about the red shift we would suppose that the number of galaxies at a distance r increases as r^2, while the contribution of each galaxy decreases as $1/r^2$. Thus the contribution to the background from the galaxies at a distance r would be independent of r. Distant galaxies would thus make an important contribution. We must therefore allow for the red shift, which weakens the contribution of distant galaxies over and above the inverse square law. A detailed calculation confirms the intuitively obvious idea that the background is approximately the same as that due to a distribution of sources without red shift that cuts off at a distance of c/H_0, at which the linear Hubble law leads to a velocity of recession equal to the velocity of light c. The background is thus roughly given by

$$\frac{nLc}{H_0},$$

where n is the number of sources per unit volume of average luminosity L (allowing, where necessary, for the cosmic evolution in luminosity). Of course a full calculation would involve knowledge of the luminosity function of the sources.

3.2.8 The Role of Horizons

An important feature of the relativistic models of the universe is that they generally contain horizons, which place limits on what is observable and also on the extent to which causal transport processes can smooth out irregularities. In the early work on relativistic cosmology these horizons were not well understood, and this led to much confusion in the literature. The situation was clarified in a basic paper by Rindler (1956). In particular, Rindler distinguished between event horizons and particle horizons. An event horizon in a cosmological model is similar to an event horizon in the space-time of a black hole or a uniformly accelerating observer in special relativity. In Rindler's words "An event horizon, for a given fundamental observer A, is a hypersurface in space-time which divides all events into two non-empty classes: those that have been, are, or will be observable by A, and those that are forever outside A's possible powers of observation." An example of a cosmological model which possesses an event horizon is the de Sitter model.

By contrast, "a particle horizon, for any given fundamental observer A and cosmic instant t_0 is a surface in the instantaneous 3-space $t = t_0$ which divides all fundamental particles into two non-empty classes: those that have already been observable by A at time t_0 and those that have not." An example of a cosmological model which possesses a particle horizon is the Einstein-de Sitter model.

In this book it is the particle horizon which will be of interest. It follows from the equation for a light ray $ds = 0$ that the proper distance to our particle horizon in models where $R(t)$ was zero a finite time ago is given by

$$l = R(t) \int_0^t \frac{cdt}{R(t)}.$$

The condition for a particle horizon to exist is then that l is finite and positive. This condition depends mainly on the behaviour of $R(t)$ for small t. In particular, if $R(t) = t^n$,

$$l = \frac{ct}{1-n},$$

and so there will be a particle horizon if $n < 1$. If $n = 2/3$

(Einstein–de Sitter model) $l = 3ct$, and if $n = 1/2$ (radiation-dominated model) $l = 2ct$.

Particle horizons are important because if they exist they limit the extent to which causal transport processes can smooth out irregularities. For example, the cosmic microwave background, which is discussed in the next section, can be smoothed out by causal transport processes only over distances less than a horizon-length. One might then expect the angular distribution of the background temperature to reveal the horizon structure of the universe, but it does not do so, since, as we shall see, it is isotropic to ~ 1 part in 10^5 (apart from a dipole term of order 10^{-3} which is believed to be due to our peculiar motion relative to the universe as a whole). Either this isotropy results from uniform initial conditions at the big bang, or one must change the behaviour of $R(t)$ for small t so as to eliminate, or greatly increase, our particle horizon, as proposed, for example, by the inflation hypothesis (Guth 1981, Linde 1990, Narlikar and Padmanhaban 1991).

3.3 The Cosmic Microwave Background Radiation

It is now well known that the universe is filled with black body radiation at a temperature $T \sim 3$ K. A Planck spectrum at this temperature peaks in the microwave region ($\lambda \sim 1$ mm). Earlier ground-based measurements, starting with the pioneering discovery of Penzias and Wilson (1965) as interpreted by Dicke *et al.* (1965), have now been powerfully supported by observations made with the satellite COBE (Cosmic Background Explorer). In particular COBE has shown (Mather *et al.* 1990) how accurately the spectrum of this background is thermal, with a temperature

$$T = 2.735 \pm 0.06 \text{ K}.$$

Observed deviations from a black body spectrum are less than 1 per cent of the peak brightness (Fig. 3.7).

An even more accurate value of the temperature has recently been obtained by COBE (Mather *et al.* 1993), namely,

$$T = 2.726 \pm 0.01 \text{ K}.$$

This is the value of the temperature which we will adopt in this book.

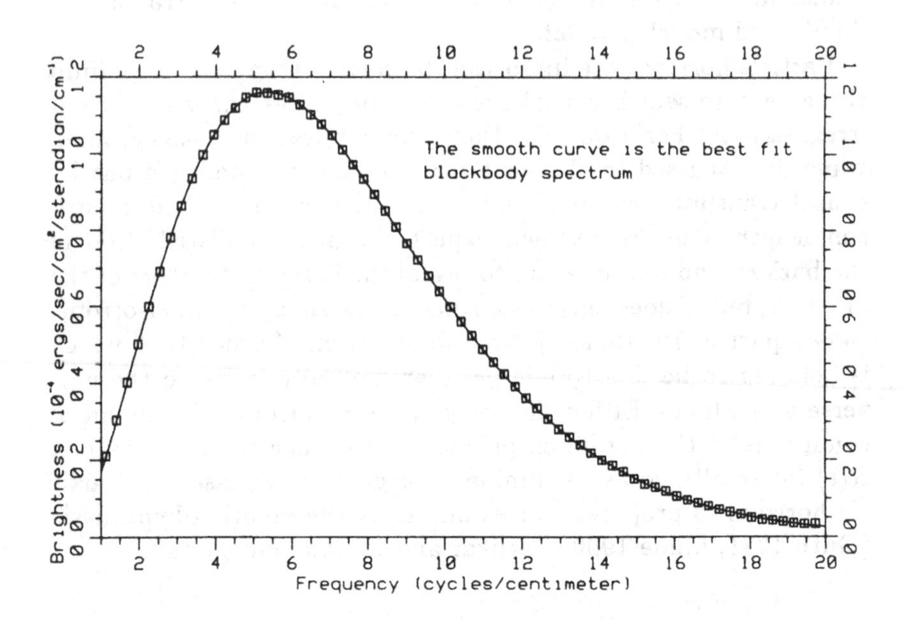

Fig. 3.7 COBE spectrum of the cosmic microwave background, compared to a black body. Boxes are measured points and show size of assumed 1% error band. The units for the vertical axis are 10^{-4} ergs cm^{-2} sec^{-1} ster^{-1} cm. [From Mather *et al.* (1990): courtesy NASA/Goddard Space Flight Center: COBE Science Working Group].

The corresponding energy density ρ_γ and the photon number density n_γ are given by

$$\rho_\gamma = 4.7 \pm 0.05 \times 10^{-34} \text{ gm cm}^{-3}$$

$$n_\gamma = 411 \pm 4 \text{ photons cm}^{-3}.$$

We will need a value for n_γ as accurate as this in our later discussion of the decaying neutrino theory.

An important property of this radiation field is its high degree of isotropy. It does, however, possess a well-established dipole variation, with $\Delta T/T \sim 10^{-3}$. This dipole is believed to arise from a net motion of the Earth relative to the universe as a whole. This motion is presumed to arise from the gravitational effect on the Earth of localised irregularities in the large scale distribution of galaxies, but the length-scale of the responsible irregularities has still to be determined.

Until recently no other anisotropy in the microwave background had been well established, the corresponding upper limit of $\Delta T/T$ lying between 2×10^{-5} and 10^{-4} on a variety of angular scales, ranging from seconds of arc to degrees. Just as this chapter was being written new COBE results were announced (Smoot *et al.* 1992) according to which the microwave background has a quadrupole moment with $\Delta T/T \sim 6\times10^{-6}$, and an angular variation on scales $\gtrsim$10 degrees of 1.1×10^{-5}. It will be important to confirm these results, because such anisotropies are expected to be produced by the gravitational action of primordial galaxies on the microwave radiation (the Sachs-Wolfe 1967 effect), and the absence of such anisotropies was beginning to constrain primordial irregularities in an embarrassing way. We will return to this question in chapter 12.

In addition to its role as a cosmological probe, the microwave background also has an important dynamical influence on the evolution of the universe. Admittedly this is not true today, since ρ_γ is much less than the mean baryon density ρ_b of the universe (which, as we shall see, is of order 10^{-31} gm cm^{-3}). However, as we go back into the past ρ_γ increases faster than ρ_b, and if we assume that the background radiation has been present since the early stages of the universe, then during these early stages the universe must have been radiation dominated. To see this, consider first the time dependence of ρ_b. In terms of the scale factor $R(t)$ we have

$$\rho_b \propto \frac{1}{R^3(t)}. \tag{14}$$

Now suppose that the Universe is filled with a homogeneous and isotropic distribution of radiation. This radiation, of course, moves with the speed of light relative to the matter, and with this motion is associated a red shift that reduces the energy of the radiation by the further factor $1/R(t)$. It is helpful here to think in terms of photons. The number of photons per unit volume decreases in the same way as the number of material particles per unit volume ($\propto 1/R^3(t)$), but whereas the rest-energy of a material particle remains the same, the energy of each photon decreases by the factor $1/R(t)$. We therefore have for the energy density ρ_{rad} of the radiation

$$\rho_{rad} \propto \frac{1}{R^4(t)}. \tag{15}$$

Perhaps an easier way of obtaining this important result is to use imaginary mirrors. Consider a small volume element of the Universe. Every time a photon leaves this element, on average a similar photon enters at the same place (we are assuming that the photons are distributed uniformly in all directions). The situation would be essentially unchanged, therefore, if we surrounded the volume element by perfectly reflecting walls that move out as the Universe expands. The red shift now arises from the Doppler effect associated with reflection at a moving mirror. The advantage of this point of view is that we have removed the complicated, and here irrelevant, cosmological aspects of the situation (the radiation at a given place having originated at a great distance, and so on). The behaviour of radiation in an expanding perfectly reflecting enclosure is well known, and is used in the standard elementary discussion of the thermodynamics of radiation (such as the derivation of Wien's law). To obtain (15) we simply use the adiabatic relation

$$pV^{\gamma} = constant,$$

where p is the radiation pressure, V the volume and γ the ratio of the specific heats (which for isotropic radiation is $4/3$). Since $V \propto R^3(t)$ we have $p \propto 1/R^4(t)$. But for isotropic radiation the pressure is related to the energy density by the equation $p = \frac{1}{3}\rho_{rad}c^2$, so we again have

$$\rho_{rad} \propto \frac{1}{R^4(t)}.$$

Now if we compare (14) and (15) we see that if there is any permanent radiation at all, then for sufficiently small $R(t)$ the energy density of radiation must have exceeded the rest-energy density of matter. In the point source models considered in this book $R(t)$ becomes arbitrarily small in the past and so we conclude that the earliest phases of these models were radiation-dominated. In these phases we have

$$R(t) \propto t^{\frac{1}{2}} \quad (t \text{ small}).$$

We are now in a position to see whether thermal equilibrium would have been set up as a result of interactions between the radiation and the matter. Since the degree of excitation of the matter was very high at sufficiently early stages (it was formally infinite at $t = 0$) we can suppose the matter to be ionised and it will suffice for our purpose to consider just the inelastic (free-free and free-bound) scattering of photons by electrons. If at any stage the scattering time was very much less than the time for the matter in the Universe to, say, halve its density, then we can be sure that thermal equilibrium would have been set up. It is a straightforward matter to show that this condition was easily satisfied at sufficiently early times.

Thus as long as there was some matter present to provide the necessary interactions we can be sure that the radiation in the early, radiation-dominated, stages of the Universe had an equilibrium (black body) spectrum completely characterised by a temperature T_{rad}. This temperature is related to the energy density ρ_{rad} by the usual black body formula

$$\rho_{rad} = aT^4_{rad},$$

where a is Stefan's constant. Accordingly $T_{rad} \propto \rho^{\frac{1}{4}}_{rad} \propto 1/R(t) \propto 1/t^{\frac{1}{2}}$. General relativity gives the precise relation

$$T_{rad} = \frac{1.5 \times 10^{10}}{t^{\frac{1}{2}}_{\text{sec}}} \text{ K (t small).} \tag{16}$$

Thus if radiation still dominated at a time of 1 second from the big bang, the radiation temperature would then have been 1.5×10^{10} K. Actually such a temperature corresponds to an energy of about 1 MeV and so electron-positron pairs would have been created and would also have been in thermal equilibrium. When allowance is made for these and also for the presence of the three types of neutrino pairs (e, μ, and τ), the total energy becomes $\frac{11}{2}aT^4$ (not $5aT^4$ despite the presence of the 5 relativistic fields: photon, electron pair, and three types of neutrino pair, because the electrons and neutrinos obey Fermi-Dirac statistics) and the temperature becomes

$$T_{rad} = \frac{2 \times 10^{10}}{t^{\frac{1}{2}}_{\text{sec}}} \text{ K (t small),}$$

a very simple result.

We must now consider what happens to the black body radiation at later times when it ceases to dominate. If there is no further interaction with matter our mirror argument assures us that the radiation would have retained its black body character, with the temperature obeying the adiabatic relation

$$T_{rad}V^{\gamma-1} = \text{constant},$$

which tells us that

$$T_{rad} \propto \frac{1}{R(t)}, \qquad (17)$$

since $\gamma = \frac{4}{3}$ and $V \propto R^3(t)$. This is in agreement with (16) for the radiation-dominated phase, but the result (17) is quite general (in the absence of appreciable interaction with matter). By contrast, the temperature of the matter (in the absence of appreciable interaction with the radiation) satisfies the adiabatic relation

$$T_{mat}V^{\gamma-1} = \text{constant},$$

where now γ for the matter (when it has become nonrelativistic) has the approximate value 5/3 which characterises a perfect gas. Thus

$$T_{mat} \propto \frac{1}{R^2(t)}.$$

We now have the following situation:

(i) The energy density of radiation decreases *more* rapidly with time than the rest-energy density of matter.

(ii) The temperature of radiation decreases *less* rapidly with time than the temperature of matter, unless the radiation and matter are strongly coupled, in which case of course they have equal temperatures. Notice that the energy density and the temperature of black body radiation determine each other, while the rest-energy density and the temperature of matter are quite separate properties.

Decoupling between radiation and matter would in fact have occurred when the temperature dropped sufficiently for the hydrogen to have become atomic. This occurred at a temperature $\sim$ 3000 K and so at a red shift $\sim$ 1000 (Peebles 1971).

We now turn to the important question of how to specify in a convenient way the relative amounts of matter and radiation in the Universe. It is not convenient to compare their energy densities because such a comparison would be time dependent. It would be preferable to have a measure that is more or less independent of time. To find such a measure we recall that the energy density of radiation falls off faster than that of matter because each photon gets red shifted as time goes on. Clearly then what we need to do is to compare the number density of photons with the number density of material particles; each decreases like $1/R^3(t)$, that is like T_{rad}^3. This last relation is a standard one for photons with a black body spectrum. Most of the energy is carried by photons whose individual energies are proportional to T_{rad}, and since the total energy density is proportional to T_{rad}^4 the number density of photons is proportional to T_{rad}^3.

3.4 Primordial Nucleosynthesis and the Baryon Density

The idea that nuclear reactions in the hot big bang might generate some of the elements was proposed in 1946 by Gamow well before the discovery (in 1965) of the cosmic microwave background. (For the early history of work on primordial nucleosynthesis see Alpher and Herman 1950 and Tayler 1990). Gamow assumed that initially matter was composed entirely of neutrons. A key step was taken in 1950 by Hayashi, who showed that at temperatures above 10^{10} K (and so at times $\lesssim$1 second after the big bang) neutrons and protons would have been in thermal equilibrium. The reason for this is that at such high temperatures electron-positron pairs would have been generated and have come into thermal equilibrium, and then the reaction

$$n + e^+ \rightarrow p + \bar{\nu}$$

would have a (weak interaction) rate faster than the expansion rate for $t \lesssim 1$ second. The reverse process would also have occurred rapidly through the presence of thermally excited anti-neutrinos.

Below 10^{10} K the weak interactions can no longer maintain the neutrons in statistical balance with the protons because the concentration of electron-pairs is beginning to drop abruptly ($m_e c^2 \sim \frac{1}{2}$ MeV $\sim \frac{1}{2} \times 10^{10}$ K). The n/p ratio is then frozen in. This frozen-in ratio, corresponding to thermal equilibrium at a temperature somewhat below 10^{10} K, is about 15 per cent. Then a series of nuclear reactions occurs, beginning with

$$n + p \rightarrow \mathrm{D} + \gamma$$

which builds up the light elements D, He^3, He^4 and Li^7. In particular most of the neutrons end up inside He^4.

Early calculations of the He^4 abundance were made by Alpher, Follin and Herman (1953), Hoyle and Tayler (1964) and Smirnov (1965). The first "modern" calculation was carried out by Peebles (1966) in which he improved the accuracy of the numerical integrations. This in turn was superseded a year later by the elaborate calculations of Wagoner, Fowler and Hoyle (1967) who took into account all the reactions known by them to occur between the light elements with the most up to date values for their cross-sections. Later work has only slightly improved on this landmark calculation, but is very elaborate in detail.

Equally elaborate is the present status of observations of light element abundances, and the corrections needed to allow for stellar processing between the early universe and the present time. It is not possible to do justice here to either the calculations or the measurements, which have been updated and reanalysed most recently by Walker *et al.* (1991), by Pagel *et al.* (1992) and by Smith *et al.* (1993). For our purposes, the main question concerns the derivation of the baryon abundance at the epoch of primordial nucleosynthesis, and to a lesser extent the derivation of the number of neutrino types.

The calculation of the primordial nuclear reactions is complicated both by the large nuclear reaction network which has to be used, and by the fact that many of the processes involved are occurring out of equilibrium. It is therefore necessary to resort to elaborate numerical computations, which makes it difficult to understand the physical issues involved, and in particular the effect of making small changes in the input data - such as the number of neutrino types. Fortunately, an analytical treatment has been recently developed by Bernstein, Brown and Feinberg (1988) which gives results correct to a few percent, and which have clarified the physics involved.

a) The abundance of He^4

It is convenient for our discussion to consider first the abundance of He^4. The primordial nuclear reactions lead to an abundance Y_4 by mass relative to hydrogen which depends on three input data, namely, the number of neutrino types N_ν, the baryon to photon ratio η, and the neutron half-life $\tau_{1/2}$. We consider these three dependencies in turn.

(i) $\boldsymbol{N_\nu}$:The value of N_ν influences the outcome of the nuclear reactions at two different stages. In both stages it does so via the neutrino contribution to the total density of the universe (each type counting as a separate relativistic species), and so, through Einstein's field equations, to the expansion time-scale. The first stage involves the fixing of the neutron-proton ratio when this ratio freezes out from thermal equilibrium. As we have seen, the idea here is that at sufficiently early epochs weak interactions keep the neutron-proton ratio in thermal equilibrium with the radiation heat bath via the reactions

$$p + e^- \leftrightarrow n + \nu_e$$

$$n + e^+ \leftrightarrow p + \bar{\nu}_e.$$

We would then have

$$\frac{n}{p} \propto e^{-\Delta m/T},$$

where Δm is the neutron-proton mass difference.

Freeze-out occurs at the temperature T_F for which the weak interaction rate starts to become less than the expansion rate. Thus the final neutron-proton ratio is given by

$$\frac{n}{p} \propto e^{-\Delta m/T_F}.$$

In fact most of these neutrons end up in He^4 nuclei as a result of the subsequent nuclear reactions. Thus the value of T_F is crucial in determining Y_4. Since this value involves the expansion rate it also involves the number of neutrino types which were relativistic at the temperature $\sim$ 1 MeV with which we are concerned here. (If there were neutrinos of mass, say, 1 GeV, they would make a negligible contribution to the density of the universe at this epoch).

After this freeze-out, the neutrinos themselves become frozen out at a temperature $\sim$ 1 MeV. Later still, electron pairs permanently annihilate (at $T \sim \frac{1}{2}$ MeV), the annihilation photons boosting the radiation bath but not the neutrinos. This leads to a reduction in the neutrino-photon ratio which changes the influence of the neutrinos on the expansion timescale. This revised timescale then affects the outcome of the nuclear capture processes which begin to be important only at this late stage. This is the second place where the value of N_ν influences Y_4.

A detailed numerical discussion of these effects is given by Bernstein, Brown and Feinberg (1988). A zero-order approximation gives $Y_4 \sim 25\%$. The final dependence on N_ν is given by

$$\Delta Y_4 = 0.012\,(N_\nu - 3).$$

Thus to determine N_ν to within one neutrino type one would need to know the primordial helium abundance to better than 4%.

It is amusing to note in passing that if the neutrinos decoupled only after the nuclear reactions have taken place, then this would correspond to an increase of 50% in the effective number of neutrino types.

(ii) $\boldsymbol{\eta}$:We next consider the dependence of Y_4 on the baryon-photon ratio η. Because this ratio turns out to be of order 10^{-10} it is usually parametrised as η_{10}, where

$$\eta_{10} = \frac{n_b}{n_\gamma} \times 10^{10}.$$

We shall see that a comparison of theory with observed abundances will lead to

$$\eta_{10} \sim 3.$$

This result is crucial for our dark matter discussion because the value of η_{10} during the nucleosynthesis epoch is also its present value, since during the adiabatic expansion in between these epochs both n_b and n_γ vary in the same way (like R^{-3}). Now today, as we have seen, $n_\gamma \sim 411 \pm 1$ cm^{-3}. Hence $n_b \sim 1.2 \times 10^{-7}$ cm^{-3} which is equivalent to

$$\Omega_b h^2 \sim 0.012.$$

The analytic discussion of Bernstein *et al.* (and the more accurate numerical computations) show that Y_4 depends only logarithmically on η_{10}. In fact we have

$$\Delta Y_4 = 0.010 \ln\,(\eta_{10}/4).$$

Thus a factor 3 uncertainty in η_{10} would give rise to an uncertainty of 1 in the derived value of N_ν.

(iii) $\boldsymbol{\tau_{1/2}}$:Finally we consider the influence of the neutron half life $\tau_{\frac{1}{2}}$. This enters our discussion because it is the most accurate determinant of the weak interaction coupling constant. Bernstein *et al.* show that, as with N_ν, it enters the analysis twice, namely in determining both the n/p ratio at freeze-out, and in the capture rates in the later nuclear reactions. It has been measured accurately only relatively recently. We shall use the value 10.33 ± 0.05 minutes obtained by Byrne *et al.* (1990). The Bernstein *et al.* analysis (and the computations) lead to

$$\Delta Y_4 = 0.014\ (\tau_{\frac{1}{2}} - 10.3).$$

Thus a 1 minute change in $\tau_{\frac{1}{2}}$ would correspond to a change in N_ν by 1 neutrino type.

We now bring this discussion together by quoting the complete expression for Y_4. This is (Walker *et al.* 1991)

$$\begin{aligned} Y_4 = 0.239 & \\ & + 0.012\ (N_\nu - 3) \\ & + 0.010 \ln\ (\eta_{10}/4) \\ & + 0.014\ (\tau_{\frac{1}{2}} - 10.3). \end{aligned}$$

In this book our main interest is in the value of η_{10}, which is poorly determined by Y_4. We therefore now consider the abundances of the other light isotopes, which can be used to give a fairly precise value for η_{10}.

b) The abundances of D, He^3 and Li^7

Considering first the deuterium abundance Y_2, the nuclear reaction calculations show that

$$Y_2 = 5 \times 10^{-4}\ \eta_{10}^{-5/3}.$$

One wants to compare this theoretical estimate with the observed deuterium abundance. The main difficulty here is that the primordial deuterium is easily destroyed later in stars, so one can obtain only a lower limit to Y_2 and so an upper limit to η_{10}. Observationally we may take $Y_2 \sim 2 \times 10^{-5}$ (this is a brutal summary of a complicated observational situation culminating in the Hubble Space Telescope observations of Linsky *et al.* 1993) and so

$$\eta_{10} \leq 7.$$

Walker *et al.* also use the fact that when D is destroyed He^3 is produced, some of which survives more easily than D itself. Thus the combined abundances Y_{2+3} of D and He^3 give an upper limit on the primordial D abundance. Adopting $Y_{2+3} < 6$ to 10×10^{-5} we have

$$\eta_{10} \geq 3 \text{ to } 4.$$

This lower limit is rather uncertain as the estimate of He^3 survival is uncertain. Putting these two limits together we have

$$3 \leq \eta_{10} \leq 7.$$

Finally we consider the abundance Y_7 of Li^7. Interestingly, this has a minimum as a function of η_{10}. If we accept the argument that population II stars have a primordial abundance of Li^7, we have $Y_7 \sim 1.2 \times 10^{-5}$, from which we derive

$$2 \leq \eta_{10} \leq 4.$$

If we now collect together the limits from D, He^3 and Li^7 we end up with

$$3 \leq \eta_{10} \leq 4.$$

The recent re-analysis by Smith *et al.* (1993) gives $2.86 \leq \eta_{10} \leq 3.77$. This is the range of values of η_{10} which we shall use in this book. It implies that $0.011 \leq \Omega_b h^2 \leq 0.015$.

c) The number of neutrino types N_ν

We now consider the cosmological derivation of the number of neutrino types N_ν. This depends on knowing Y_4. There is a long history of measurements of helium abundances in different objects, and of extrapolations back to the primordial abundance, allowing for the production of He^4 in stars. Detailed discussions are given by Davidson, Kinman and Friedman (1989), Pagel and Simonson (1989), Torres-Peimbert, Peimbert and Fierro (1989), and most comprehensively by Pagel *et al.* (1992) and Pagel and Kazlauskas (1992). The last two references give

$$Y_4 = 0.228 \pm 0.005$$

or

$$Y_4 < 0.242 \quad \text{(95 per cent confidence limit)}.$$

The significance of this result for N_ν depends on a thorough understanding of the error bars involved. Pagel *et al.* (1992) concluded that

$$2 < N_\nu \leq 3.2.$$

This result (or one close to it) was first obtained when the laboratory limit on N_ν was considerably greater, so that the cosmological limit represented a definite prediction. This prediction has now been verified by observations of Z_0 particles at SLAC and CERN. A recent CERN result is (Dydak 1991)

$$N_\nu = 3.01 \pm 0.10.$$

This agreement represents a triumphant verification of the standard model of the hot big bang, and has led physicists generally to accept the predictions of this model, although it may still be subject to small variations if inhomogeneities associated with the quark-hadron transition at a temperature $\sim$ 200 MeV turn out to be important (Bonometto and Pantano 1993, Mathews, Schramm and Meyer 1993).

It should be remembered, however, that the laboratory accelerators are not measuring exactly the same quantity as the primordial nucleosynthesis argument. For example, supersymmetry particles like the photino, which are not coupled to the Z_0, would not be counted by the accelerators, but could contribute to the nucleosynthesis count if they were relativistic at that epoch. On the other hand, a neutrino of mass $\sim$ 1 GeV, with couplings similar to those of the e, μ and τ neutrinos, would be excluded by the accelerator data, but not by the nucleosynthesis count. This is an important point because such neutrinos, now excluded by the Z_0 data, were once a serious candidate to be the dark matter.

3.5 The Density, Age and Expansion Rate of the Universe

We come now to the question which is crucial for the existence of dark matter in the universe as a whole, namely, what is the total mean density ρ of the universe and how does it compare with estimates of its baryon density ρ_b?

Measurements of ρ usually involve the expansion rate as measured by the Hubble constant H_0, so we shall consider here also the measured value of H_0. These measurements lead to a test of the Robertson-Walker models since these models relate ρ and H_0 uniquely to the age t_0 of the universe (in the absence of a cosmological constant). We therefore also consider here direct astrophysical measurements of the age t_0 of the universe.

Unfortunately the attempts which have been made to measure ρ, H_0 and t_0 are surrounded by controversy, and the procedures involved are elaborate and for the most part highly technical. Whole books and conferences have been devoted to the measurement of each one of these parameters. It would not be possible to do justice to these discussions in this short book, but we do need to know the range within which each parameter is likely to lie. I shall therefore confine myself to quoting the generally agreed range, while recognising that individual practitioners might have strong preferences for a narrower range. I also give references to some recent technical discussions of these questions.

The density ρ is usually derived from some form of cosmic virial theorem applied to the observed peculiar velocity field related to galaxies, clusters or superclusters of galaxies. Since distance estimates and so H_0 are involved in the analysis, one usually derives a value for the density parameter $\Omega = \rho/\rho_{crit}$, rather than for ρ itself. Some of the methods measure only the contribution to Ω of the mass which follows the light, and so would not include a dark smooth substratum in which we are here primarily interested. Other methods measure either the total Ω or sometimes $\Omega^{0.6}/b$, where the bias parameter b is the ratio of the r.m.s. fluctuations in the light and the mass.

There seems to be general agreement that

$$\Omega \geq 0.1 \text{ to } 0.2,$$

Peebles (1986), Trimble (1987), Salucci, Persic and Borgani (1992), with several authors finding

$$\Omega \sim 1$$

(e.g. Kaiser *et al.* 1991, Dekel 1991, Frenk 1991, Heavens 1991).

One could also determine Ω by measuring the deceleration parameter q_0 and then using the relation $q_0 = \frac{1}{2}\Omega$. Recently Kellermann (1993) obtained a value for q_0 from VLBI observations of the angular size-red shift relation for compact radio sources, which he believes may be free of significant evolutionary effects. Kellermann's relation is shown in Fig. 3.8. He obtained $q_0 \sim \frac{1}{2}$, and so $\Omega \sim 1$. This interesting result still needs to be confirmed.

If Ω were exactly 1 at the present epoch it would be exactly 1 at all epochs (Einstein-de Sitter model). By contrast, if Ω differs from

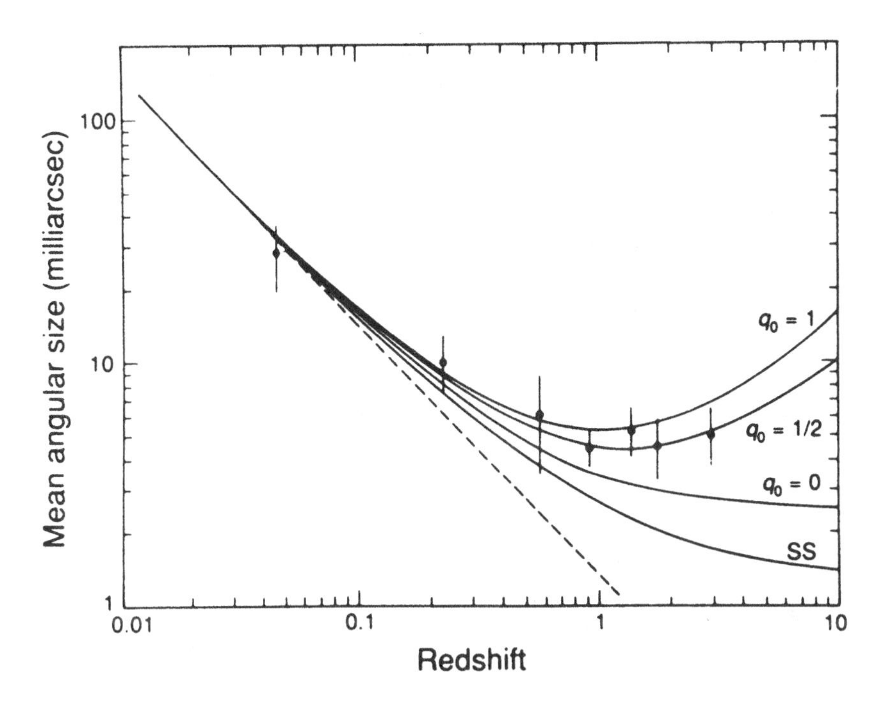

Fig 3.8 Mean angular size against redshift for 82 compact sources. From left to right, individual points are based on 8, 4, 11, 13, 20, 13 and 10 sources, respectively. The error bars represent the spread in angular size for each redshift bin. The solid curves represent the expected dependence for Friedmann cosmological models for a standard source with a component separation of 41 pc and deceleration parameters q_0 of 0, $\frac{1}{2}$ and 1 and the steady state model (SS); the dashed line shows the $1/z$ law observed for the separation of double-lobed extended sources. [Reprinted with permission from *Nature* (From Kellermann 1993)].

1, it is close to 1 only for a restricted range of epochs. This has led some theorists to propose that Ω is indeed exactly 1. Another point of view is to suppose that Ω is driven close to 1 for an extended period by some dynamical process e.g. inflation (Guth 1981, Linde 1990, Narlikar and Padmanhaban 1991). In this book we shall regard Ω as a quantity to be determined empirically.

We now wish to compare our estimates of Ω with the derived baryon density of the universe. This density, as determined from considerations of primordial nucleosynthesis, is given directly as

Table 3.1. H_0 *determinations*

Authors	H_0	
Birkinshaw (1991)	37 ± 9	
Rhee (1991)	50 ± 17	
Salucci and Sciama (1991)	54 ± 7	
Soucail and Fort (1991)	75 ± 20	$(q_0 = 1/2)$
Soucail and Fort (1991)	92 ± 26	$(q_0 = 0)$
Branch (1992)	61 ± 10	
Branch and Tammann (1992)	56 ± 6	
Jacoby *et al.* (1992)	73 ± 11	
Pierce *et al.* (1992)	86 ± 12	
Press *et al.* (1992)	< 65	
Schmidt *et al.* (1992)	60 ± 10	
Tammann (1992)	55 ± 2	
Salucci, Frenk and Persic (1992)	61 ± 5	
van den Bergh (1992 a, b)	76 ± 9	
Sandage *et al.* (1992)	51 ± 9	
Sandage (1993a)	43 ± 11	
Sandage (1993b)	45 ± 12	
Branch and Miller (1993)	51 ± 12	

ρ_b, and is independent of H_0. This is equivalent to having derived $\Omega_b h^2$. Thus to make a comparison we need to know the value of H_0.

The problem of measuring H_0 has led to the most famous and the most violent controversy in the history of astronomy and cosmology. Good surveys with ample references have been given by Rowan-Robinson (1985) and van den Bergh (1989, 1992 a, b). These surveys can be summarised by saying that H_0 probably lies in the range 50 to 100 km sec^{-1} Mpc^{-1}. The most recent determinations are given in table 3.1.

One problem with the measurement of H_0 which has only recently been clearly recognised is that the velocity field of nearby galaxies contains significant deviations from a pure Hubble flow. Thus one cannot obtain a reliable (and consistent) value for the Hubble constant by simply comparing the red shifts of those galaxies with their estimated distances (however accurately derived). One either needs to determine the local peculiar velocity field from a dynamical model, or to confine oneself to galaxies so distant that their peculiar velocity is negligible.

In view of the wide variation in the results still being obtained we shall, for the purposes of this book, continue to adopt the conventional range

$$H_0 = 50 \text{ to } 100 \text{ km sec}^{-1} \text{ Mpc}^{-1}$$

so that

$$0.5 \leq h \leq 1.$$

We now return to the question of the baryon density. We saw on page 54 that the primordial nucleosynthesis arguments lead to

$$0.011 \leq \Omega_b h^2 \leq 0.015.$$

Hence

$$0.01 \leq \Omega_b \leq 0.06.$$

By contrast, $\Omega \geq 0.1$. Thus, although the gap between Ω_b and Ω is not large (unless the proponents of $\Omega \sim 1$ are correct), we conclude that, on these figures, the universe contains a significant component of non-baryonic dark matter. This component could even be the dominant one.

Finally we consider the age of the universe t_0, which has been extensively discussed by many authors and is reviewed in Vangioni-Flam *et al.* (1990). The observed age is mainly derived either from nuclear cosmochronology (using unstable nuclei with a half-life comparable to the age of the universe (e.g. Fowler 1987)), or from fitting evolved globular cluster stars to theoretical evolutionary tracks in the Hertzsprung-Russell diagram. The latter method is usually considered to be the better one, but it suffers from a number of uncertainties, including the distances to the stars concerned, and the roles of the oxygen abundance and of helium diffusion in these stars. Recent discussions have been given by Profitt and Michaud (1991), Profitt and Vandenberg (1991), Renzini (1991), Carney *et al.* (1992), Chaboyer *et al.* (1992 a, b), Walker (1992), Vandenberg (1992) and Dearborn *et al.* (1992). These discussions cover the range

$$t_0 = 12 \text{ to } 20 \times 10^9 \text{ years}.$$

This result has immediate implications for our cosmological modelling. The Robertson-Walker models (with $\lambda = 0$) imply that

$$t_0 = f(\Omega)/H_0,$$

where

$$f(\Omega) = \frac{2}{3} \qquad \Omega = 1$$

$$f(\Omega) = (1-\Omega)^{-1} - \frac{\Omega}{2}(1-\Omega)^{-\frac{3}{2}} \cosh^{-1}(\frac{2}{\Omega} - 1) \qquad \Omega < 1$$

$$f(\Omega) = \frac{\Omega}{2}(\Omega - 1)^{-\frac{3}{2}}\left[\cos^{-1}(\frac{2}{\Omega} - 1) - \frac{2}{\Omega}(\Omega - 1)^{\frac{1}{2}}\right] \qquad \Omega > 1$$

Consider first the simple case of $\Omega = 1$ (corresponding to the Einstein-de Sitter model). If $h = 1$, we would have $t_0 = 6.6 \times 10^9$ years, which is definitely too low. If $h = 0.5$, we would have $t_0 = 13 \times 10^9$ years, which is acceptable (although on the low side according to some astronomers). By contrast, if $\Omega \ll 1$ we would have $t_0 \sim 1/H_0$. For $h = 1$ we would then have $t_0 = 10 \times 10^9$ years which still seems too low. For $h = 0.5$ we would have $t_0 = 20 \times 10^9$ years which is acceptable (especially if one allows for the time taken to form the globular cluster stars in the first place).

There is an interesting conflict underlying these figures, which springs from a tendency amongst various experts to push (independently) the values of all three quantities Ω, H_0 and t_0 to the higher ends of their ranges. This would be compatible with general relativity only if one introduces a non-zero cosmological constant λ, whose desirability is otherwise itself controversial. My own view is tied up with the neutrino decay theory which is discussed in part 3 of this book. We will find that for this theory to be valid (with $\lambda = 0$) we will need

$$\Omega \sim 1$$

$$h \sim 0.56$$

$$t_0 \sim 12 \times 10^9 \text{ years.}$$

These values (just) lie within the presently allowed observational ranges.

4
The Identity of the Dark Matter

4.1 Introduction

A good starting-point for a discussion of the identity of the dark matter is to ask how much of it is baryonic and how much is non-baryonic, both in individual objects and in the universe as a whole. If we accept the standard model of the hot big bang then we require that

$$0.011 \leq \Omega_b h^2 \leq 0.015.$$

If in addition we accept that $\frac{1}{2} \leq h \leq 1$, then

$$0.01 \leq \Omega_b \leq 0.06.$$

Let us compare this range of values for Ω_b with the mean density Ω_{vis} due to visible baryons in galaxies and clusters of galaxies. This quantity has been estimated a number of times over the years. We shall follow a recent re-evaluation by Persic and Salucci (1992), who arrived at a lower value than in earlier estimates. These authors use recent detailed results on the mass to light ratios of galaxies, on the galaxy luminosity function, and on the x-ray properties of rich clusters of galaxies. They then find that

$$\Omega_{vis} \sim (2 + 0.8\, h^{-1.3}) \times 10^{-3},$$

the first term coming from single galaxies and the second from clusters. According to this estimate

$$\Omega_{vis} < \Omega_b.$$

We next compare Ω_b with estimates of Ω itself. These estimates were discussed at the end of the last chapter where we saw that

$$\Omega \geq 0.1.$$

We may therefore conclude that

$$\Omega_{vis} < \Omega_b < \Omega.$$

It follows from these inequalities that there must exist both baryonic and non-baryonic dark matter in the universe. We shall therefore need to locate both these components of the dark matter. We may find in particular that their relative proportions are different in different types of object, depending on their modes of formation. Such cosmogonical questions are not well understood at the moment, but we can attempt to determine empirically the relative proportions in different objects by direct astrophysical means.

One example of this empirical approach is based on a study of the ratio of the dark to visible density as a function of position in an object. We saw in chapter 2 that this ratio decreases rapidly with distance from the centre in some rich clusters of galaxies, but increases rapidly in normal spirals . It is natural to relate this difference in behaviour to the likely difference in the dissipational properties of baryonic and non-baryonic matter. For example, Eyles *et al.* (1991), who discovered the high central condensation of the dark matter in the Perseus cluster from their precise x-ray data, suggested that this concentration probably showed the dark matter in the cluster to be dissipative, and therefore baryonic. As we shall see, most forms of non-baryonic dark matter would be weakly interacting and so effectively non-dissipative. It is then natural to suggest that most of the dark matter in normal spirals could be non-baryonic, although it does not follow that this must be so.

4.2 Baryonic Dark Matter

Baryonic dark matter could take a number of different forms, such as faint stars, jupiters, black holes and intergalactic gas. A detailed discussion of the first three possibilities can be found in Carr (1990), Lynden-Bell and Gilmore (1990), Ashman (1992) and Silk (1992). The last possibility is discussed in chapter 7. A specific proposal has been made by Thomas and Fabian (1990), who suggested that cooling flows in rich clusters of galaxies end up as low mass faint stars. These flows are widely, but not universally, believed to arise in the inner regions of many x-ray emitting clusters, where the hot intracluster gas which produces the x-rays is dense enough for its cooling time to be less than the age of the clusters. This gas is then compressed by the pressure of the outlying gas which retains its high temperature. A recent survey of these cooling flows has been given by Fabian, Nulsen and Canizares (1991). The proposal of Thomas and Fabian would fit in well with the discovery that the dark matter in rich clusters of galaxies is more strongly concentrated towards the centre of the clusters than is the visible matter in hot gas and galaxies.

Cooling flows may also have occurred in some form at much earlier cosmological epochs, and it has been suggested by Thomas and Fabian (1990), Ashman and Carr (1988, 1991), Ashman (1990) and White and Frenk (1991) that as a result the halos of galaxies also consist of low mass stars formed in such flows. The existence of such low mass stars has been studied recently by Daly and McLaughlin (1992), Richstone *et al.* (1992) and Richer and Fahlman (1992) who concluded that the question is still unresolved. A direct search for such stars, and for black holes, is currently being conducted by attempting to observe the gravitational microlensing which would be produced by these objects (Paczynski 1986a,b, Griest 1991).

4.3 Non-Baryonic Dark Matter

Non-baryonic dark matter is usually thought to consist either of weakly interacting massive particles (wimps) or topological defects in a gauge field (Kolb and Turner 1990, Sarkar 1991). The wimps owe their existence in the universe today to a double process. First they were pair created in the hot big bang and rapidly reached thermal equilibrium with the ambient heat bath, their annihilations being continually balanced by further pair creations. Then later their annihilation and creation rates became smaller than the expansion rate of the universe, and the particle distribution decoupled from the heat bath at a temperature T_D.

A crucial distinction emerges at this point, according to whether the particles were relativistic ($kT_D > m_0c^2$) or non-relativistic ($kT_D < m_0c^2$) at decoupling. In the former case one speaks of hot dark matter (HDM) and in the latter case of cold dark matter (CDM). This distinction plays a crucial role in the calculation of the particle density which survives to the present day, and also in the processes of galaxy formation.

The survival of the HDM pairs is easy to calculate. At decoupling they were still essentially as abundant as the photons in the heat bath (a more precise statement will be made later). Thereafter the number density of the now non-interacting particles continued to decrease as the inverse cube of the scale factor of the universe, just as did the number density of the photons in the heat bath. As a result, the number density n of the particles which survive to the present day is of the same order as the number density of the photons in the 3 K microwave background. (Later on we will supply a precise factor relating the two number densities).

It follows from this that if the particles had zero rest-mass their energy density today would be of the same order as that of the 3 K background, namely $\sim 10^{-34}$ gm cm^{-3}, which is only $\sim 10^{-5}$ of the critical density. They would therefore make a negligible contribution to Ω. In addition, such zero mass particles could not be confined to the halos of galaxies.

The situation would be completely different if the particles have a non-zero rest-mass. During the expansion of the universe the momentum of each particle decreases inversely as the scale factor of the universe (just like the momentum of a photon). When the momentum of a particle was red shifted below $m_0 c$, the particle became non relativistic. This would have occurred before the present day if $m_0 c^2$ exceeds the mean photon energy in the 3 K background, that is $\sim 10^{-3}$ eV. Then the contribution ρ_p of the particles to the present density would be $n m_0$. In this case we would have the fundamental result

$$\Omega_p \propto m_0.$$

We could thus have $\Omega_p \sim \Omega$ if m_0 had the appropriate value. More generally this argument leads to an upper limit on m_0, since Ω cannot greatly exceed unity.

The corresponding calculation for the survival of CDM is more complicated because the particle density at decoupling was then heavily suppressed by the Boltzmann factor $\exp[-m_0/kT_D]$. One has to use the Boltzmann equation to track the annihilations at temperatures close to T_D (this calculation, with references to the original work, is discussed by Bernstein 1988 and Kolb and Turner 1990). One finds that the present mass density $n m_0$ in these particles is a decreasing function of m_0, since the reduction imposed by the Boltzmann factor is exponential in m_0, which dominates the linear increase from the factor m_0. In this case the condition $\Omega = O(1)$ leads to a lower limit on m_0. If the particles have a neutrino-like annihilation cross-section one finds that

$$m_0 \gtrsim 2 \text{ GeV}.$$

This calculation breaks down if m_0 is very much larger than 2 GeV. In this regime ρ_p increases as m_0^2, and the bound on Ω then leads to

$$m_0 < 10^5 \text{ GeV}.$$

Massive neutrinos with $m_0 \sim 2$ GeV are no longer viable candidates for the dark matter since LEP has shown that there are only 3 neutrino-types whose mass is less than half that of the Z_0, that is, less than 46 GeV. In addition Mori *et al.* (1992) used results from Kamiokande to exclude neutrinos in the mass range 6 to several hundred GeV as the dark matter in the halo of our Galaxy. However, other speculative candidates for CDM exist such as supersymmetry particles and axions. These possibilities are discussed by Kolb and Turner (1990) and Sarkar (1991), and attempts to detect them in the laboratory are described by Smith and Lewin (1990). More recent discussions have been given by Griest and Roszkowski (1992) and Urban *et al.* (1992) amongst others.

The literature contains extensive discussions of the different roles played by HDM and CDM in the processes of galaxy formation, and in the resulting theoretical small and large-scale distribution of galaxies. Some authors have attempted to discriminate between the various theories by comparing their predictions with the rapidly accumulating observations. In my opinion galaxy formation is still too ill-understood a set of processes to be used in this way. I will therefore not describe these discussions in this book, and simply refer the reader to the extensive literature.

There is, however, one important respect in which HDM does have an advantage over CDM , and that is that at least neutrinos are known to exist. Moreover, although it is still not known experimentally whether they have a non-zero rest-mass, it would be a quite natural theoretical result of elementary particle physics (unlike the case of photons for which gauge invariance would forbid a non-zero rest-mass). Finally the most attractive solution of the solar neutrino problem — the MSW effect or one of its variants — would require neutrino types to convert into one another as they propagate (Bahcall 1989). This would require them to have a non-zero rest-mass. I therefore take the point of view that massive neutrinos are the most conservative — and most likely — candidate for the non-baryonic dark matter. For this reason I give in the next section a detailed account of their properties as a dark matter candidate. The remainder of the book is devoted to a more speculative possibility — namely, that these dark matter neutrinos decay into hydrogen (and nitrogen) ionising photons at a rate which would make them the dominating ionising source both in individual galaxies (outside HII regions) and in the universe as a whole. The resulting theory is highly constrained and would solve a number of astrophysical and cosmological problems involving ionised material.

4.4 Massive Neutrinos as Dark Matter Candidates

The role of massive neutrinos in cosmology was emphasised twenty years ago by Cowsik and McClelland (1972) and by Marx and Szalay (1972), following earlier work by Gerstein and Zeldovich (1966). Some of the ideas involved have close links with the role of neutrinos in the sun and in supernovae. These links are particularly important because neutrinos have been detected coming from the sun and from supernova 1987A in the Large Magellanic Cloud. Detailed discussions of these questions, and of particle physics theories of neutrinos, may be found in a number of recent books, for example, Bahcall (1989), Kolb and Turner (1990), Börner (1990), Winter (1991), Mohapatra and Pal (1991) and Boehm and Vogler (1992).

This material is very extensive, but for our purposes it is sufficient to note that neutrinos are weakly interacting spin $\frac{1}{2}$ particles of three types only, namely ν_e, ν_μ and ν_τ. It is still not known whether they have non-zero rest-masses. The following upper limits on their rest-masses have been obtained from laboratory data

(Holzschuh 1992, Albrecht *et al.* 1992)

$$m_{\nu_e} < 9 \text{ eV},$$

$$m_{\nu_\mu} < 250 \text{ keV},$$

$$m_{\nu_\tau} < 31 \text{ MeV}.$$

There is also a model-dependent upper limit on m_{ν_e} of 15 to 30 eV derived from the neutrino observations of supernova 1987A.

As we shall see, one of the key results of the cosmological discussion is that, for stable neutrinos, the observation that Ω cannot greatly exceed unity implies that (Gerstein and Zeldovich 1966)

$$\sum_{\nu_i} m_{\nu_i} \lesssim 100\ h^2 \text{ eV},$$

the sum being over the three neutrino types. This constraint is much stronger than the laboratory ones for ν_μ and ν_τ. We shall also see that if a neutrino type ν_i provides most of the dark matter in our Galaxy, then Liouville's theorem implies that

$$m_{\nu_i} \gtrsim 27 \text{ eV}.$$

Thus the hypothesis that the dark matter in our Galaxy consists mainly of neutrinos would automatically imply that neutrinos make a significant contribution to Ω, and could readily lead to $\Omega = 1$.

4.5 Relic Neutrinos from the Hot Big Bang

Neutrinos of each type are pair produced in the hot big bang by the weak interaction process

$$\nu + \bar{\nu} \leftrightarrow e^- + e^+.$$

They will come into thermal equilibrium with the radiation heat bath if the weak interaction collision rate t_c^{-1} exceeds the expansion rate t_e^{-1} of the universe. The expansion rate in seconds is given by

$$t_e^{-1} = \frac{1}{4}\left(\frac{T}{10^{10}} \text{ K}\right)^2$$

as we saw on page 47. The collision rate is of order

$$t_c^{-1} \sim n_\nu \sigma c,$$

where σ is the weak interaction cross-section. Since $\sigma \sim G_F^2 E^2 \sim G_F^2 T^2$ and $n_\nu \sim T^3$ (in units in which $\hbar = c = k = 1$) we have

$$t_c^{-1} \sim G_F^2 T^5.$$

We then find that the neutrinos are in thermal equilibrium with the heat bath if

$$T \gtrsim 1 \text{ MeV}.$$

Thus the decoupling temperature T_D of the last section is in this case ~ 1 MeV. Since the rest-masses of ν_e and ν_μ are less than 1 MeV, they were relativistic at decoupling and so constitute hot dark matter. The same is true for ν_τ if it is stable on the time scale of the age of the universe.

The evolution of the neutrino distribution function after decoupling is determined by the simple consideration that the momentum of each neutrino is inversely proportional to the scale factor of the universe. So long as the neutrinos remain relativistic their distribution function conforms closely to that of a Fermi-Dirac gas of zero rest-mass particles at a temperature which itself varies inversely as the scale factor of the universe. In this sense the decoupling does not appreciably change the thermal behaviour of the neutrinos until their temperature dropped down to $\sim m_\nu$. Then the neutrinos moved into the non-relativistic regime, and their distribution function became a non-equilibrium one. Its behaviour can still be easily determined from the fact that the momentum of each neutrino continues to vary inversely as the scale factor of the universe. In particular, one finds that in this non-relativistic regime their root mean square velocity v_ν at a red shift z is given by

$$v_\nu = 6\left(\frac{30 \text{ eV}}{\text{m}_\nu}\right)(1+z) \text{ km sec}^{-1}$$

(Bond, Efstathiou and Silk 1980).

We now come to the crucial question of determining the number density of neutrinos both at the epoch of primordial nucleosynthesis and at the present time. As was first realised by Alpher, Follin and Herman (1953) (see also Peebles 1971, Weinberg 1972) although this number density will clearly be of the same order as the number density of photons in the heat bath, the precise relation is affected by the fact that at $T \sim m_e \sim \frac{1}{2}$ MeV the electron pairs began to annihilate permanently. Their annihilation photons would have boosted the temperature of the heat bath, but not the neutrino temperature, since the neutrinos decoupled from the heat bath a little earlier (at $T \sim 1$ MeV). (Actually this decoupling is not quite complete at $T \sim \frac{1}{2}$ MeV, and the slight resulting heating of the neutrinos by annihilating electron pairs would increase the final density of neutrinos by 1.5% for ν_μ and ν_τ (Dodelson and Turner 1992)).

This photon boosting was computed by Alpher, Follin and Herman, who argued that it is a good approximation to treat the permanent annihilation of the electron pairs as an adiabatic process. Then the entropy of the electron pairs becomes added to that of the photons, thereby boosting their number density and temperature. We now calculate the amount of this boosting.

To do this we need to know the entropy density of a Fermi-Dirac and a Bose-Einstein gas. This entropy density is proportional to ρ/T, where ρ is the energy density of the gas. In order to derive n_ν from T_ν and n_γ from T_γ we will also need to know the number density of a Fermi-Dirac and a Bose-Einstein gas. These densities involve integrals over the Fermi-Dirac and Bose-Einstein distribution functions, the ones for the number density involving a transcendental function (the Riemann ζ function).

Let us first relate to one another n and ρ for Fermi-Dirac and Bose-Einstein gases. We can do this without carrying out the integrals involved by using a simple algebraic trick.† Consider the expression $n_{B-E} - n_{F-D}$. By combining the denominators $e^{p/T} - 1$ and $e^{p/T} + 1$ in the integrands for these quantities, we immediately see that

$$n_{B-E} - n_{F-D} = \frac{1}{4} n_{B-E},$$

so that

$$n_{F-D} = \frac{3}{4} n_{B-E}.$$

By a similar argument we find that

$$\rho_{F-D} = \frac{7}{8} \rho_{B-E}.$$

† I am grateful to M. J. Rees for telling me of this trick.

Since a Dirac electron has 4 states whereas a photon has only 2, one finds that when the electron pairs annihilate adiabatically the photon number density increases by the factor $1 + 7/4 = 11/4$. In order to relate n_ν to n_γ after the annihilation we must also allow for the factor $3/4$ connecting the number densities of Fermi-Dirac and Bose-Einstein gases. We therefore arrive at the crucially important result

$$n_\nu = \frac{3}{11} n_\gamma,$$

which holds at all times after decoupling. This result is independent of both the Hubble constant and the neutrino mass (as long as the neutrinos were relativistic at decoupling). It also holds for both Majorana and Dirac neutrinos, since the right handed massive Dirac neutrinos would be expected to be very weakly interacting and so would decouple much earlier than left handed neutrinos. Accordingly their number density would be suppressed by the permanent annihilation of many particle species at $T \sim 200$ MeV (Shapiro *et al.* 1980, Dolgov and Zeldovich 1981, Olive *et al.* 1981).

These considerations were implicitly used in the previous chapter in connexion with the influence of the number of neutrino types N_ν on the outcome of primordial nucleosynthesis. We now discuss their implications for the neutrino density at the current epoch. Let us first consider n_γ. Now we need to carry out the appropriate integral explicitly. This is a standard problem and one finds that

$$\begin{aligned} n_\gamma &= \frac{8\pi}{h^3} \int_0^\infty \frac{p^2\, dp}{e^{pc/kT} - 1} \\ &= \pi\zeta(3) \left(\frac{kT}{hc}\right)^3, \end{aligned}$$

where here h is Planck's constant and $\zeta(3) \sim 1.2$. Inserting the present value of the microwave background temperature

$$T = 2.726 \pm 0.01 \text{ K},$$

we find that

$$n_\gamma = 411 \pm 4 \text{ cm}^{-3}.$$

Hence

$$n_\nu = 112 \pm 1 \text{ cm}^{-3}$$

for each neutrino type.

The present density ρ_{ν_i} in each neutrino type is then given by $n_\nu m_{\nu_i}$, which is independent of the Hubble constant. We thus find that

$$m_{\nu_i} = 94\ \Omega_{\nu_i} h^2\ \text{eV}$$

for each neutrino type (apart from the small correction for the incomplete decoupling of the neutrinos at $T \sim \frac{1}{2}$ MeV). Assuming that $\Omega_\nu < 1.1$ then gives us the upper limit on $\sum_i m_{\nu_i}$ quoted earlier. Alternatively, if we assume that $\Omega_{\nu_x} \sim 1$ for the most massive neutrino type ν_x (recalling that $\Omega_b \leq 0.015\ h^{-2}$), we would have

$$m_{\nu_x} \sim 94\ h^2\ \text{eV}.$$

If h is close to the lower end of its permitted range (as would be required to solve the age problem of the universe), we would then have

$$m_{\nu_x} \sim 30\ \text{eV}.$$

4.6 Massive Neutrinos in Galactic Halos

Since neutrinos are non-dissipative it seems reasonable to consider them as possible candidates for the dark matter in the outer halos of galaxies. Following Cowsik and McClelland (1973) it is useful to begin by constructing a zero order model in which the neutrinos constitute a self-gravitating completely degenerate system. Such systems have been discussed in detail by Landau and Lifshitz (1958) who showed that their mass M and radius R obey the relation

$$MR^3 = \frac{91.9\hbar^6}{G^3 m_\nu^8}.$$

Cowsik and McClelland modelled the Coma cluster in this way, using a relatively low value of m_ν ($\sim$ 10 eV). For our later discussion we are more interested in the possibility that $m_\nu \sim 30$ eV. For this choice we would have

$$MR^3 \sim 8 \times 10^{11} (20\ \text{kpc})^3\ \text{M}_\odot.$$

So for $R \sim 20$ kpc, we would have $M \sim 8 \times 10^{11}\ \text{M}_\odot$, which makes a reasonable model of a galaxy like our own.

We would not in fact expect a galaxy formed from relic neutrinos to be completely degenerate, and Tremaine and Gunn (1979) have provided us with a different model based on known properties of these relic neutrinos. They pointed out that after decoupling the neutrinos are effectively free of all interactions except gravitational ones. They then applied Liouville's theorem and argued that the maximum value of the phase space density of the neutrinos in a galaxy could not exceed the maximum value in the distribution function at decoupling, which occurs at zero momentum. However, if each macroscopic region of the galaxy contains neutrinos coming from different regions of the original distribution function, then the resulting coarse-grained phase space density would be everywhere less than the maximum possible value. This question of phase mixing has been much discussed in the literature, without a clear-cut consensus emerging. There is some evidence from n-body simulations that the phase space density at the centre of a galaxy is close to its maximum value (Melott 1982 a, Carlberg 1986), and for simplicity we tentatively adopt this hypothesis here.

We follow the convenient formulation of Peebles (1980), which is itself based directly on the work of Tremaine and Gunn. We write for the initial distribution function of the neutrinos

$$df = \frac{4}{h^3} d^3p d^3r N(r,p),$$

where here h is Planck's constant and the occupation number $N \leq \frac{1}{4}$ for left-handed neutrinos. Following Caldwell and Ostriker (1981) we adopt for the halo of our Galaxy the simple model

$$N = \frac{\exp(-3v^2/2v_0^2)}{4(1 + r^2/a^2)},$$

where the core radius $a \sim 8$ kpc and the neutrinos are assumed to have an isotropic and isothermal velocity distribution. In practice the velocity distribution may be anisotropic and r dependent, but since we will find that $m_\nu \propto v_0^{-1/4}$, this uncertainty may not be too important.

Since the neutrinos are now non-relativistic, their momentum is mv and their root mean square velocity is v_0. Integrating over momenta at $r \gg a$, and assuming that the neutrinos provide most of the mass in the halo, we obtain Peebles' results for $r \gg a$

$$\rho = \frac{v_0^2}{6\pi G r^2}$$

and

$$m_\nu^4 = \frac{1}{6\pi}\left(\frac{3}{2\pi}\right)^{3/2}\frac{h^3}{Gv_0 a^2}.$$

This result should be more reliable than the one based on the cold degenerate model, since the kinetic energy of the neutrinos at the sun's position is about 4.5 times greater than their Fermi energy. If we drop the assumption that the phase space density at the centre of the Galaxy has its maximum possible value, this last equation would give a lower limit for m_ν.

Inserting values for our Galaxy we have $a \sim 8$ kpc and $v_0 \sim 230$ km sec^{-1} (Salucci and Frenk 1989). We then obtain

$$m_\nu = 27.6 \pm 1 \text{ eV},$$

the uncertainty following from reasonable estimates of the uncertainties in v_0 and a. It is remarkable that this value is so close to our previous value for m_ν, derived from the assumptions that neutrinos provide essentially the critical density, and that the Hubble constant h is close to the lower end of its permitted range (as would be required to solve the age problem of the universe).

However, the matter is not that simple. Smaller galaxies, especially dwarfs, contain dark matter but have much smaller values of v_0, and although we may not know their core radii in neutrinos, it seems likely that the same analysis would lead to an appreciably larger value of m_ν, which would not be compatible with the $\Omega = O(1)$ constraint. This problem has been much discussed in the literature and attempts have been made to resolve it by resorting to rather complicated distribution functions for the neutrinos. Gerhard and Spergel (1992) have recently argued strongly against these attempts for Draco and Ursa Minor. If one assumes that these dwarf galaxies are dominated by 30 eV neutrinos they would require very large core radii (~ 10 kpc) and masses ($\sim 4 \times 10^{11}$ $M_\odot$). This would make their dynamical friction decay times in the Galactic halo significantly shorter than a Hubble time. Moreover these limits turn out to be insensitive to assumptions about the anisotropy of the neutrino distribution.

It may therefore be necessary to assume that the dark matter in dwarf galaxies is baryonic, although it may be non-baryonic in normal spirals. We recall in this connexion that because there exists both baryonic and non-baryonic dark matter, we have in any case to determine for each individual type of object what are the relative proportions of the two types of dark matter. There would then have to be some cosmogonical reason why this ratio is greater for smaller galaxies, just as there would have to be a cosmogonical reason why the ratio is greater for at least some rich clusters of galaxies, than it is for normal spirals.

We could even regard these requirements as providing us with clues to the still little-understood physics and dynamics underlying galaxy formation. These are problems for the future.

Part Two

Ionisation Problems in Astronomy and Cosmology

5
Diffuse Ionisation in the Milky Way

5.1 Introduction

The interstellar medium of our Galaxy contains a widespread component of ionised gas with fairly well-determined average properties. The source of the ionisation has puzzled astrophysicists for many years (e.g. Mathis 1986, Kulkarni and Heiles 1987, Reynolds 1991, 1992, Walterbos 1991, Heiles 1991). There are five major problems which contribute to the mystery. They are the following:

(i) The scale height of the ionised gas $\sim$ 670 pc (Nordgren *et al.* 1992), whereas the sources usually considered (e.g. ionising radiation from O stars or supernovae) have a much smaller scale height ($\sim$ 100 pc).

(ii) The power requirements which the sources must satisfy in order to maintain the ionisation are rather large and probably rule out any conventional source except O stars (Reynolds 1990b).

(iii) The interstellar medium is normally regarded as being highly opaque to hydrogen-ionising radiation, so that it is not clear how this radiation can travel hundreds of parsecs from the parent O stars to produce the diffuse ionised gas (Mathis 1986, Reynolds 1984, 1987, 1992, Heiles 1991).

(iv) The same opacity problem arises when one studies in detail (Reynolds 1990a) the ionisation along the line segments to two pulsars with accurately known parallactic distances.

(v) The mean electron density in opaque intercloud regions within a few hundred parsecs of the sun is remarkably constant in different directions (Reynolds 1990a, Sciama 1990c). If the opaque gas has a sufficiently tortuous distribution to explain problem (iii), it is surprising that the resultant electron density is so uniform.

In this chapter we shall elaborate on these problems. As we shall see in chapter 9, all of them would be immediately solved if the ionising source consists of photons emitted by decaying dark matter neutrinos in the Galaxy.

5.2 The Scale Height of the Ionised Gas

An early attempt to estimate the scale height of the ionised gas in the Galaxy was made by Bridle and Venugopal (1969). They used a variety of phenomena involving the electron density to derive a lower limit of 400 pc for this scale height. They also pointed out problem (i) in the final sentence of their paper: "An electron disk of this thickness is unlikely to be supported by photoionization due to hot stars, as these are largely confined to a disk of thickness only 100 pc in the z direction." They opted instead for ionization by cosmic rays, which was a popular theory at that time, but which was later ruled out, as we shall see.

Other authors, using newer data, arrived at similarly large scale heights. Falgarone and Lequeux (1973) obtained a lower limit of 500 pc, Readhead and Duffett-Smith (1975) derived an actual value of 670 pc, while Cordes, Weisberg and Boriakoff (1985) gave a lower limit of 500 pc.

More recently a definitive result has been provided by Reynolds (1989a, 1991). He used new data on the dispersion of pulsar signals which results from their propagation through an ionised medium. The observed dispersion immediately gives an accurate value for the total column-density of free electrons $\int_0^l n_e ds$ between the observer and each pulsar. This so-called dispersion measure (DM) has recently become available for several pulsars in high latitude globular clusters which are so distant that they lie outside the ionised layer in the interstellar medium. Reynolds was therefore able to derive the asymptotic value for the z component of DM, which he found to be about 23 cm^{-3} pc (Fig. 5.1).

Since earlier studies of DM for nearer pulsars had shown that in the galactic plane $\bar{n}_e = \int n_e ds / \int ds$ is about 0.025 cm^{-3} (e.g. Weisberg, Rankin and Boriakoff 1980), he concluded that the scale-height of the free electrons is about 900 pc. (His earlier result (Reynolds 1989a) of 1.5 kpc was based in part on erroneous DM's for pulsars in the globular cluster 47 Tuc). Because of this result, and of other important studies which he has made of the ionised layer, this region is now called the Reynolds layer. Its scale

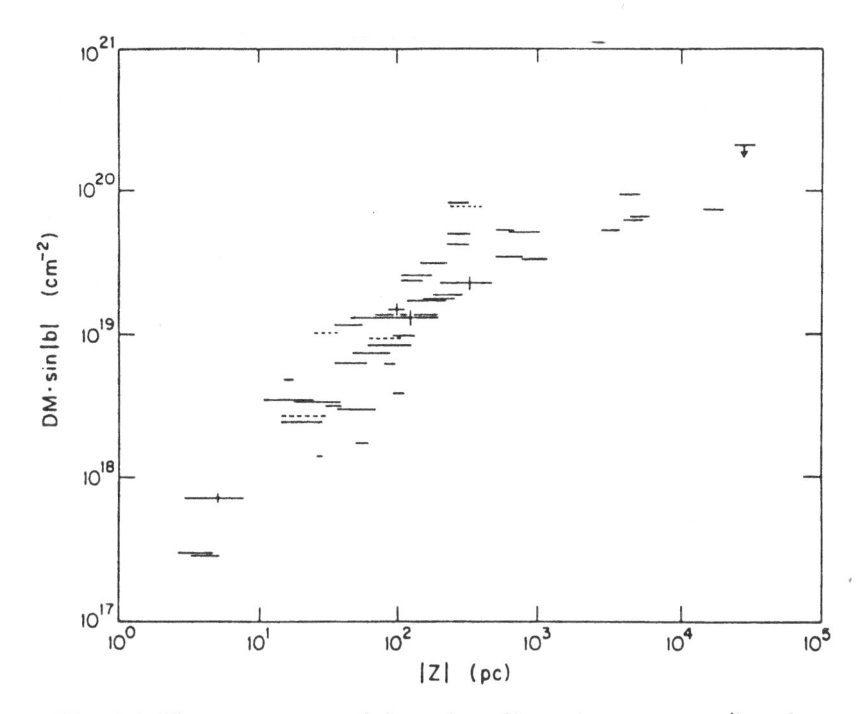

Fig. 5.1 The component of the pulsar dispersion measure (i.e. the column density of n_e) perpendicular to the Galactic plane is plotted against distance $|z|$ from the midplane. Pulsars behind the Gum nebula are indicated by dashed lines; the pulsar in the Large Magellanic Cloud is indicated by an upper limit arrow. [From Reynolds 1991].

height has recently been revised down to 670(+170, −140) pc by Nordgren *et al.* (1992), from their new value for $\bar{n}_e$ of $0.033 + 0.002$ cm^{-3}.

5.3 The Power Requirements of the Reynolds Layer

These power requirements have been discussed by Reynolds (1990b). Following earlier studies of the energetics of the interstellar medium (e.g. Dalgarno and McCray 1972, Spitzer 1978, Kulkarni and Heiles 1987, Cowie 1987 and Black 1987), Reynolds distinguished between the power required to maintain the ionisation and that required to maintain the temperature of the gas.

The sources of power must overcome the energy lost through hydrogen recombination and collisional cooling, since it is likely that the Reynolds layer is in a steady state. The cooling rate is strongly temperature dependent, and various analyses (based on observed emission-line ratios (Reynolds 1989b)) suggest that outside cold compact clouds most of the gas is at a temperature within a factor 2 of 10^4 K.

We first consider the processes responsible for maintaining the gas at a temperature $\sim 10^4$ K. These processes have been extensively discussed in the references just given (and more recently in Reynolds and Cox 1992 and Sciama 1993b), and we cannot enter here into this aspect of the interstellar medium. Suffice it to say that the main heating agent has not yet been identified with certainty, but that a likely candidate is the photoemission of electrons from interstellar dust grains or macromolecules. The photons responsible for this emission would mainly have an energy below the hydrogen ionisation threshold at 13.6 eV, would come from ordinary stars, and could easily penetrate the neutral hydrogen in the interstellar medium.

Of more direct importance for us in this book is the power required by the process mainly responsible for maintaining the ionisation of the Reynolds layer. This power can be estimated from Reynolds' (1984, 1989c) observations of the $H\alpha$ emission from this layer (Fig. 5.2). The $H\alpha$ intensity is usually measured in rayleighs R ($1\ R = 10^6$ photons cm^{-2} sec^{-1}). Reynolds found that at Galactic latitudes $|b| > 5$ degrees the $H\alpha$ intensities I_α generally decrease with increasing $|b|$ in a manner consistent with $I_\alpha \sin|b| = 1\ R$.

This cosecant law for I_α is attributed to the recombining gas having a slablike distribution. It follows that the recombination rate along a vertical line of sight from the Earth on one side of the galactic plane is about 2×10^6 cm^{-2} sec^{-1} (as only about half the recombinations lead to the emission of $H\alpha$ photons). Since each recombination involves the net emission of at least 13.6 eV, the minimum power required to maintain the ionisation on both sides of the plane is about 4×10^{-5} ergs cm^{-2} sec^{-1}. As Reynolds (1990b) pointed out, this minimum power is about 50 percent of the total power produced by supernovae, and about 10 percent of the ionising power produced by O and B stars (Bregman and

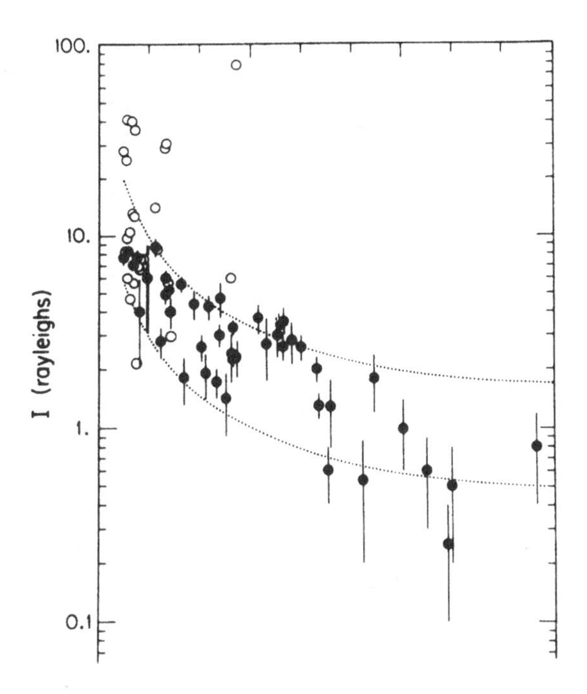

Fig. 5.2 The $H\alpha$ intensity in Rayleighs (R) plotted against the galactic latitude of the observation direction. The open circles represent directions toward or within 5 degrees of emission nebulosity visible on photographic surveys. The two dashed curves represent the expected variation of intensity with galactic latitude for a uniformly emitting galactic disk with $I \sin |b| = 0.5R$ and $1.7R$ and no extinction. $1\ R = 2.42 \times 10^{-7}$ ergs cm^{-2} sec^{-1} ster^{-1} at $H\alpha$. [From Reynolds 1984].

Harrington 1986). As we have seen, the problem with this latter source is how the ionising photons are able to propagate to where they are needed.

An alternative possible ionising source which was once much discussed is low energy cosmic rays, which were assumed to propagate unimpeded throughout the interstellar medium. Of course, if the cosmic rays are produced by supernovae, as is widely believed, we have the same power problem all over again. However, we can also make a direct attack on this question. It is helpful for this purpose to work in terms of the required cosmic ray ionisation rate per hydrogen atom ζ. Since the total ionisation rate in a vertical direction on one side of the galactic plane is 2×10^6 cm^{-2} sec^{-1} and since the column density of neutral hydrogen is about 3×10^{20} cm^{-2} (see below), we require $\zeta \sim 7 \times 10^{-15}$ sec^{-1}. However, the known cosmic ray flux would lead to a much smaller value of ζ, namely 7×10^{-18} sec^{-1} (Spitzer and Tomasko 1968). Since low energy cosmic rays are magnetically excluded from the solar system, it was suggested (Spitzer and Tomasko 1968, Field, Goldsmith and Habing 1969) that a large excess flux of cosmic rays with an energy of about 2 MeV per nucleon might pervade the Galaxy. This could bring ζ up to the required value.

Doubt was cast on this possibility in 1973, when Meszaros showed that CIII would then be overproduced. It was rather directly disproved in 1976 when Shaver measured ζ in individual interstellar clouds. He did this by combining low frequency hydrogen recombination line observations with HI absorption measurements. In this way he obtained $\zeta < 2 \times 10^{-16}$ sec^{-1}. Subsequently Shaver, Pedlar and Davis (1976) obtained $\zeta \leq 3 \times 10^{-17}$ sec^{-1}, while Payne, Salpeter and Terzian (1984) found $\zeta < 4$ to 16×10^{-16} sec^{-1}.

One can also determine ζ from observations of molecular abundances in the clouds, as has been done by Black and Dalgarno (1973), O'Donnell and Watson (1974), Jura (1974) and Glassgold and Langer (1974). These authors obtained values for ζ which generally lie in the range 10^{-17} to 10^{-16} sec^{-1}, but for some clouds $\zeta < 10^{-17}$ sec^{-1}.

We thus arrive at the important conclusion that the ionising agency for the Reynolds layer cannot penetrate interstellar clouds. Since in addition this agency is unlikely to be shock waves, which would not be expected to produce the observed rather uniform ionisation (Mathis 1986), the widely held view is that photoionisation by ultra-violet radiation is responsible. This hypothesis was originally introduced by Silk and Werner (1970) and by Werner, Silk and Rees (1970).

5.4 The Opacity of the Interstellar Medium

The opacity of the interstellar medium for hydrogen-ionising photons is mainly due to neutral hydrogen HI. The distribution of

HI is very complicated, and has recently been reviewed by Dickey and Lockman (1990). It is not possible to do justice to their discussion here, but fortunately our main interest is in the large scale vertical structure, which can be fairly simply summarised. Their "best estimate" for this structure near the sun is a combination of two Gaussians, of central densities 0.4 and 0.1 cm^{-3} and full width half maxima of 210 and 530 pc, and an exponential with $n(0) = 0.06$ cm^{-3} and a scale height of 400 pc. The distribution has a FWHM of 230 pc, a central density of 0.57 cm^{-3}, and a column density of 6×10^{20} cm^{-2} through a full disk.

To see the implications of this distribution for the opacity of the HI gas, we note that the photoionisation cross-section for HI at threshold is 6.8×10^{-18} cm^2. Thus a threshold photon with energy 13.6 eV would have a mean free path of only ~ 0.05 pc in an HI gas of density 1 cm^{-3}.

To overcome this opacity one would have to suppose that the HI distribution is full of holes. This is the point of view of Norman and Ikeuchi (1989) and of Norman (1991) who consider the effect on the interstellar medium of multiple-supernova events occurring in the disk. They argue that the resulting explosions create essentially vacant chimneys connecting the disk to the halo, and they relate these chimneys to the vertical "worms" observed in HI (Koo *et al.* 1992). Ionising photons from O stars could then propagate through these chimneys to great heights.

It is not clear whether such a model would produce the observed diffuse distribution of free electrons. For example, Mathis (1986) has estimated that about 10% of the sky as seen from a typical O star would need to have a column density of HI $< 10^{17}$ cm^{-2}. Kulkarni and Heiles (1987) regard this as an unreasonable requirement within the portion of the cloud layer that is at heights less than 100 pc, where they consider the O stars to be ineffective. However, the pulsar DM data tell us that we need diffuse ionisation at these low heights, a point to which we shall return in the next section.

Outside the cloud layer there are runaway O stars to consider, and also planetary nebula nuclei and hot white dwarfs (Nordgren *et al.* 1992). However, there remains the opacity problem and the true ionisation source is still widely regarded as a mystery (e.g. Heiles 1991, Reynolds 1991, 1992). We shall see in the next chapter that the same problem exists for the diffuse ionisation observed in other galaxies such as NGC 891.

5.5 The Diffuse Ionised Gas near the Sun

In a paper significantly entitled "What ionises the interstellar hydrogen towards PSR 0950+08 and PSR 0823+26?" Reynolds (1990a) considered carefully the origin of the ionisation along the lines of sight to these two pulsars. He chose these pulsars for special study because they were the only ones at the time with accurately known distances, obtained from parallax measurements using VLBI techniques (Gwinn *et al.* 1986). When combined with the dispersion measures of these pulsars, the distances then gave accurate values for the mean electron density along the well-defined line segments to the two pulsars. Armed with this information Reynolds then examined in detail all the O and B stars and hot white dwarfs near these line segments to see whether they were capable of producing the ionisation. He concluded that the B stars and white dwarfs could not be the ionisation sources. Moreover the nearest O stars with Lyman continuum luminosities capable of producing the ionisation are approximately 300 to 400 pc from the line segments. The interstellar medium would normally be regarded as being highly opaque to ionising radiation over much smaller distances. One would need to invoke the existence of very extended regions or channels within which the gas has a low density and is predominantly ionised.

We give a few details of Reynolds' discussion here, but the interested reader should consult the original paper for a full account of all the candidate stars. PSR 0950+18 has galactic co-ordinates $l = 229°$, $b = +44°$, a distance 127 ± 13 pc, and the column density of free electrons towards it is 9.2×10^{18} cm^{-2}. The corresponding quantities for PSR 0823+26 are (197°, +32°; 357 ± 80 pc; 6.0×10^{19} cm^{-2}). These line segments are separated by about 28 degrees and have their far end points at distances above the galactic plane of +90 pc and +190 pc, respectively. They do not intersect any known HII regions. The corresponding mean volume electron densities are 0.023 cm^{-3} and 0.054 cm^{-3}, respectively.

Reynolds pointed out that the local electron density along a line of sight is not necessarily approximately constant. In particular one should take into account the fact that the sun is immersed in a bubble of size $\sim$ 100 pc containing x-ray emitting gas at a temperature $\sim 10^6$ K and a corresponding low density $\sim 5 \times 10^{-3}$ cm^{-3} (Cox and Reynolds 1987).

Towards PSR 0950+08 this cavity appears to extend out to a distance of about 80 pc. The ionised gas responsible for the pulsar dispersion measure must then be confined primarily to the outer 47 pc of the PSR 0950+08 line segment and have an electron density ~ 0.056 cm^{-3}. The hot cavity then accounts for only about 10% of the total electron column density. By contrast, the correction for the other pulsar is small. Thus essentially the same value for n_e in normal HI regions was derived for both line segments, namely ~ 0.05 cm^{-3}. It is this value which Reynolds was unable to ascribe to nearby O stars unless the HI distribution is full of holes.

After Reynolds wrote his paper another pulsar parallax was determined (Bailes *et al.* 1990). The relevant data are: PSR 1451-68 (315°, −8.5°; 450 $\pm$ 60 pc; 2.6×10^{19} cm^{-2}). Thus the mean electron density along the line segment to this pulsar is 0.019 cm^{-3}. Possible variations in this quantity along the line of sight were considered by Sciama (1990c). Again the local bubble is unimportant in this case (the path length $\sim$ 50 pc). However, the line segment passes right through Loop I, a well-known nearby supernova remnant. The position and geometry of this Loop are clearly shown in Fig. 5.3, taken from Cox and Reynolds (1987).

The data indicate that one should take a path length $\sim$ 300 pc through the Loop and an electron density $\sim 10^{-2}$ cm^{-3} (Iwan 1980, Nousek *et al.* 1982). This low density is associated with the high temperature of this x-ray emitting region. The contribution of Loop I to the dispersion measure is therefore ~ 3 cm^{-3} pc. As the observed dispersion measure is 8.6 cm^{-3} pc, the final $\sim$ 100 pc of path length through a normal HI region contributes ~ 5.6 cm^{-3} pc. Thus the mean electron density in this section of the path ~ 0.056 cm^{-3}.

It is rather striking that we arrive at the same mean electron density in the opaque regions of each line segment, namely 0.05 cm^{-3}. If the ionisation is to be ascribed in each case to relatively distant O stars whose ionising photons propagate through holes in the HI distribution, one would expect to find significant

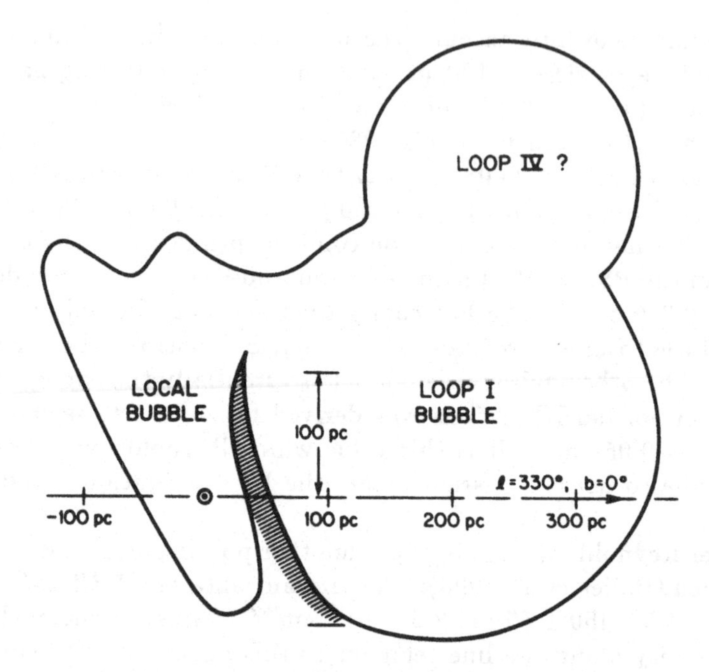

Fig. 5.3 Schematic representation of the relationship between the Local and Loop I Bubbles and the Sun, including the intervening wall of HI. [From Cox and Reynolds 1987: Reproduced, with permission, from the *Annual Review of Astronomy and Astrophysics*, Vol. 25, © 1987 by Annual Reviews Inc].

variations in the resulting electron densities in different directions. We could avoid both the opacity problem and the problem of the constant electron density if, to quote Reynolds, there is "a source of ionisation that is more smoothly distributed within the galactic disk than O stars and that has 10% to 15% of the total ionising power of the O stars." This is precisely what is achieved in the decaying neutrino theory.

5.6 The Ionisation of the Local HI Cloud

The sun is known to be immersed in a cloud of HI density $\sim 0.1\ \text{cm}^{-3}$, column density $N(HI) \sim 10^{18}\ \text{cm}^{-2}$, radius ~ 5 pc and temperature $\sim 10^4$ K. This region is the central part of what is now called the Local Interstellar Medium (LISM). Following the review article of Cox and Reynolds (1987) we take the LISM to consist of material within the nearest 10^{19} cm^{-2} in both HI and HII. The distance to that column density is large, ranging from roughly 30 pc in the galactic plane to as much as 200 pc within 20 degrees of the North Galactic Pole.

Within that region, and outside the local HI cloud, there is an irregularly shaped volume containing a hot x-ray emitting gas with a temperature $\sim 10^6$ K, a density $\sim 5 \times 10^{-3}$ cm^{-3} and a mean radius ~ 100 pc — the Local Bubble, which has already been referred to in the previous section.

There have been many observational studies of the ionisation state of the LISM using both back-scattered solar ultra-violet radiation and the absorption spectra of nearby stars. The situation is still somewhat confused; useful surveys can be found in Paresce (1984), Cox and Reynolds (1987), and Frisch and York (1987). Various observations suggest that the hydrogen in the local cloud is partially ionised. These observations include the solar backscatter data, the spectrum of the white dwarf HZ 43, and the magnesium column densities towards some stars. According to Cox and Reynolds the ionisation fraction probably lies between 0.3 and 0.7. This would be in accord with the electron density of 0.05 cm^{-3} derived above for the HI regions towards the three pulsars with known parallaxes.

A detailed calculation of the expected level of hydrogen and helium ionisation in the local cloud has been made recently by Slavin (1989) and by Cheng and Bruhweiler (1990). They included all the sources of ionisation known to them, namely B stars, white dwarfs, the hot gas in the Local Bubble, and the interface between the local cloud and the Local Bubble. Their results have recently been tested by observations of HI and HeI in the spectrum of the white dwarf G191-B2B by the Hopkins Ultraviolet Telescope which flew on the satellite Astro I in December 1990 (Kimble *et al.* 1993). This star is about 47 pc away, and its extreme ultra-violet spectrum had been previously observed by Green *et al.* (1990), who detected absorption by HeI as well as HI. The more accurate observations of Kimble *et al.* lead to an abundance ratio HI/HeI $= 11.6 \pm 1$ in the local cloud.

The closeness of this ratio to the present value of H/He in the Galaxy (~ 10) means that H and He are nearly equally ionised in the local cloud. This fits in well with the model calculations of Slavin and of Cheng and Bruhweiler, but differs from earlier measurements, which had indicated that HI/HeI ~ 6. This latter result (and other considerations) had led Cox and Reynolds (1987) to suggest that an additional source of hydrogen ionisation may be present, possibly fossil ionisation from a past supernova. While the need for extra ionisation would fit in well with the earlier considerations in this chapter, the measurements from HUT are in fact compatible with the lower end of the range of hydrogen ionisation inferred by Cox and Reynolds ($n_e \sim 0.03$ cm^{-3}). There is thus no need to invoke an extra source of ionisation although one might be present. This point has been stressed by Kimble *et al.* , and will be further discussed in chapter 9.

We now consider the ionisation level of nitrogen in the LISM - this will be important for our neutrino decay theory. The ratio of ionisation cross-section to recombination rate for nitrogen is within a factor of 2 of that of hydrogen, while its ionisation potential (14.5 eV) is close to that of hydrogen (13.6 eV). Accordingly we would expect these two elements to be ionised to approximately the same degree wherever they are observed. This expectation is realised, for example, in the Orion HII nebula, where the column densities of NI and HI are observed to satisfy $\log \mathrm{NI/HI} \sim -4.2$ (Lequeux *et al.* 1979), which is within a factor 2 of the cosmic abundance ratio of N and H. It is also realised in mainly neutral interstellar clouds, where $\log \mathrm{NI/HI}$ is again observed to be close to -4.2 (Ferlet 1981, York *et al.* 1983).

A value for NI has been measured in the LISM by Gry *et al.* (1985). They observed the line of sight towards the star βCMa, which they found to be "exceptionally vacant". Although this star is 200 pc away, the only HI cloud they observed along the line of sight was the local cloud, which in this direction has an HI column density of 1 to 2.2×10^{18} cm^{-2}, and extends over a few (~ 5) pc. By contrast, the HII column density is estimated to be $\sim 10^{19}$ cm^{-2}, and to fill 40 to 90 pc of the line of sight.

Gry *et al.* found that the column density of N in this direction is 6 to 8.2×10^{13} cm^{-2}. This is in agreement with the standard value of -4.2 for $\log \mathrm{NI/HI}$. It would seem from this that whatever is ionising hydrogen is also ionising nitrogen.

5.7 The Widespread Ionisation of Nitrogen in the Interstellar Medium

The aim of this section is to describe the recent evidence that nitrogen has a once-ionised component extending throughout the interstellar medium. Since the ionisation potential of nitrogen slightly exceeds that of hydrogen, similar opacity and scale height problems arise. For this reason the prevalence of NII will be important for the neutrino decay theory. The evidence comes from the observation of forbidden NII lines at 122 and 205 microns by the Far Infra-Red Absolute Spectrophotometer (FIRAS) on the Cosmic Background Explorer (COBE) (Wright *et al.* 1991). Earlier evidence for the widespread ionisation of nitrogen in the interstellar medium came from observations of optical emission lines of NII by Reynolds *et al.* (1977) and Sivan *et al.* (1986) which were further analysed by Mathis (1986). However, these lines were observed in the galactic plane and in the direction of the Orion-Eridanus loop, which lies near an association of O and B stars and supernovae.

By contrast, FIRAS observed the 205 μ line in a large variety of directions in the Galaxy, including many at high latitudes (Fig. 5.4). Towards the poles this flux is observed to be 7 ± 2 ergs cm^{-2} sec^{-1}.

The significance of this flux has been discussed by Wright *et al.* (1991), Sciama (1992) and Reynolds (1992). The upshot of this discussion is again that the nitrogen must be ionised wherever the hydrogen is ionised.

As we shall see later, the same statement holds good for the edge-on spiral galaxy NGC 891. Fig. 6.1 shows that the distribution of NII tracks that of HII out to several kiloparsecs from the plane (Keppel *et al.* 1991). We conclude that if photons emitted by decaying neutrinos are the main ionisation source for hydrogen in the interstellar medium and in NGC 891, then they must also be the main ionisation source for nitrogen in these regions. This will turn out to be a crucial feature of the neutrino decay theory.

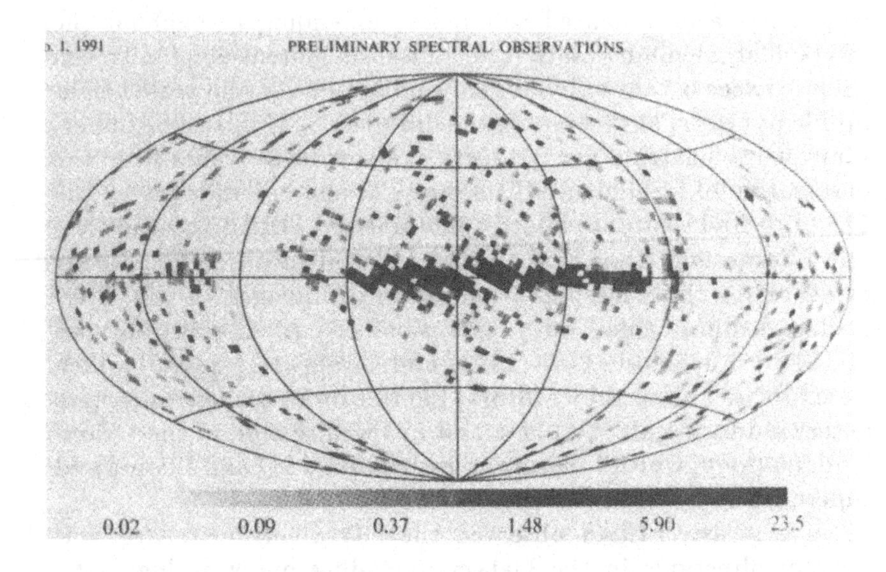

Fig. 5.4 Grey-scale representation of [NII] line flux at 205 microns. Units on the scale bar are 10^{-6} ergs cm^{-2} sec^{-1} $ster^{-1}$. [From Wright *et al.* 1991].

5.8 Ultraviolet Radiation inside Dark Interstellar Clouds

This section takes its title from that of a review article by Sorrell (1992). Studies of the abundances of various atoms and molecules in dark interstellar clouds have indicated that there may exist a greater flux of far uv radiation in the interior of these clouds than would be expected on the basis of their high opacity. In particular the abundance ratio of neutral carbon to CO in these regions is anomalously high. This has become known as the C^0/CO ratio problem. Equilibrium models (Langer 1976, Tielens and Hollenbach 1985 a, b) predict a value for this ratio which is smaller than the observed one by a factor $\sim 10^{-5}$. This anomaly could be explained if the deep interiors of the clouds contain excess far uv radiation capable of dissociating CO.

According to Sorrel the possible sources of this excess uv radiation are (1) decaying dark matter (a possibility which we discuss in chapter 9), (2) excess penetration of ambient starlight into cloud interiors and (3) the Prasad-Tarafdar (1983) mechanism in which cosmic rays induce H_2 luminescence in the Lyman-Werner bands.

In order to calculate the penetration of starlight one needs to have a good model for the dust grains which provide most of the opacity in the far uv. This is a complicated problem. As Sorrel points out, one needs to know the size distribution of the grains, the size dependence of their optical properties, their spatial distribution and the effects arising from small grains sticking to large ones. This last point, the sweeping up of small grains by large ones, is particularly important because it is the small grains which are mainly responsible for the far uv opacity. Given all these complications one cannot tell definitively whether the C^0/CO ratio problem can be solved along these lines. Sorrel's conclusion is that it can.

In the Prasad-Tarafdar mechanism cosmic rays with energies $\sim$ 10 to 100 MeV ionise H_2 and release secondary electrons with average energies $\sim$ 30 to 36 eV. These electrons then collisionally excite electronic states of H_2. When these excited states decay they emit uv radiation in the Lyman-Werner bands. This radiation then dissociates the CO via line absorption into predissociated bound states. The efficacy of this mechanism has been evaluated by Gredel, Lepp and Dalgarno (1987), who concluded that it could provide a factor $\sim 10^3$ out of the needed $\sim 10^5$, in order to account for the observed C^0/CO ratio. They suggested that the remaining factor $\sim$ 100 could be obtained by appealing to a differential depletion of C and O due to adsorption on grains, and to a velocity spread of the CO molecules of $\sim$ 10 km sec^{-1} which would increase the number of wavelength coincidences between the uv radiation and the absorption lines.

It thus does seem to be possible to solve the C^0/CO ratio problem in a number of ways, but at the cost of introducing a variety of complicating factors which are difficult to evaluate precisely. By contrast, we shall see in chapter 9 that the neutrino decay theory could solve the problem in a straightforward and direct manner (Tarafdar 1991). Moreover this solution, in conjunction with the hypothesis that the decay photons can ionise nitrogen, itself leads to a very delicate test of the theory, and would constrain the energy of the decay photons with a precision of one part in 200.

5.9 New Data on Diffuse Ionisation in the Milky Way

We end this chapter by considering the recent paper by Spitzer and Fitzpatrick (1993) entitled "Composition of Interstellar Clouds in the Disk and Halo I.HD 93521." This paper is of great importance for our understanding of the diffuse ionisation in the Milky Way, and provides strong support for the neutrino decay theory of this ionisation, as we shall discuss in chapter 9.

Spitzer and Fitzpatrick used the Hubble Space Telescope to observe the ultraviolet absorption spectrum of the Galactic halo star HD 93521, which lies 1500 pc above the Galactic plane (Spitzer and Fitzpatrick 1992). This spectrum was already known to contain nine clearly resolved components. Five of these components have large velocities relative to the local standard of rest, ranging from -27.4 to -64.6 km $\sec^{-1}$. The remaining four components have lower relative velocities, ranging from -16.5 to 9.0 km $\sec^{-1}$. From the 21 cm data also available for the components (Danly *et al.* 1992) one knows that each of them is highly opaque to Lyman continuum radiation, and that the temperature of each component ~ 6000 K.

Spitzer and Fitzpatrick were able to derive the average electron density n_e in each of these components by considering collisional excitation of the upper ($J = \frac{3}{2}$) fine-structure level of the C^+ ground state, whose column density in each component they were able to measure. By comparing the values of n_e and n_{HI} in the components with their global values in the Galaxy in warm occupied regions, they concluded that most of the neutral and ionised hydrogen along this line of sight are clumped together into the observed components. Moreover they argued that if each component were assumed to contain two regions, with HI concentrated in one region and free electrons and protons in the other (as in the model of McKee and Ostriker 1977) several difficulties of interpretation would appear. They concluded that such a model seems implausible but cannot be excluded.

This analysis provides the first fairly clear-cut evidence that the free electrons associated with H α emission and pulsar dispersion measures are probably located in warm opaque regions. Thus even if ionised tunnels exist in the interstellar medium permitting hydrogen-ionising stellar photons to propagate considerable distances from their parent stars, these photons would be unable to penetrate components of the type discussed by Spitzer and Fitzpatrick. The difficulties discussed at the beginning of this chapter are thereby strongly reinforced.

The values of n_e for each component derived by Spitzer and Fitzpatrick are very interesting. They derived these values from the equilibrium equation

$$A_{12}n(C^{+*}) = \gamma_{12}n_e n(C^+),$$

where γ_{12} is the rate co-efficient for electron excitation and A_{12} is the Einstein co-efficient for the transition to the ground state. Unfortunately $n(C^+)$ was not measured and so they estimated it from their observed value of the column density of S^+ in each component, assuming that

(i) column densities can replace particle densities,
(ii) the depletion of S due to adsorption on dust grains is negligible,
(iii) the depletion of C is by a factor of 2.

The use of (i) means that the derived value of n_e is a weighted average through each component. Assumption (ii) is based on their observed ratio of $N(S^+)$ to $N(H^0)$ in each component. Assumption (iii) is described as plausible, but lacks supporting observational evidence along the line of sight. We shall question this assumption below.

In this way Spitzer and Fitzpatrick derived for the slowly moving components an average n_e of 0.11 cm^{-3} with a 1σ dispersion of only 0.01 cm^{-3}. For the faster components n_e is systematically less, averaging 0.04 cm^{-3}, and shows much larger relative scatter about this mean. Since these components are moving fast enough to sustain shock waves, which could easily destroy the equilibrium conditions assumed above and introduce the larger scatter in n_e, we prefer to concentrate here on the results for the more quiescent components.

For these components the derived values of n_e are remarkably constant (0.10 ± 0.01, 0.11 ± 0.007, 0.11 ± 0.008 and 0.12 ± 0.015 cm^{-3}). This reminds us of our analysis of the lines of sight to the three nearby pulsars in section 5.5, where we also found constant values for n_e(0.054, 0.056 and 0.056 cm^{-3}). We now notice that there is just a factor of 2 between these sets of values. Since the first set depends on the depletion factor of C, which was arbitrarily taken to be $\frac{1}{2}$, it is natural to suggest that the depletion of C may, like that of S, be negligible in these components. This would also make it easier to understand the remarkable constancy of n_e in the four quiescent components. With this assumption the values of n_e in the slow components would become 0.05, 0.055, 0.055 and 0.06 cm^{-3} and all seven occupied regions would have the same electron density. Whether this constancy holds more widely could be tested by using HST to observe quiescent warm absorbing regions along the lines of sight to other stars. In chapter 9 we shall see that this constancy is predicted by the neutrino decay theory for such regions lying within 1 to 2 kpc of the sun.

6
Diffuse Ionisation in Spiral Galaxies

6.1 Introduction

We saw in the last chapter that the Milky Way contains diffuse ionised gas (DIG) with a large scale height. We also saw that there is strong, but not decisive, evidence that conventional sources in the Galaxy are not adequate to account for the observed ionisation. What seem to be needed are sources which are smoothly distributed, so that the opacity of the neutral hydrogen can be overcome, and which possess a large enough scale height to account for the large scale height of the DIG. Dark matter neutrinos in the Galaxy would be expected to possess both these properties, as we discuss in detail in the next part of this book. If the radiative decay of these neutrinos is to be a serious candidate for the ionisation source of the DIG in our Galaxy, we would expect to find the same ionisation problems in nearby galaxies whose structure is similar to ours. This is the subject of the present chapter.

There is one advantage and one disadvantage in studying the ionisation in other galaxies. The advantage is that by observing from a point outside the galaxies it is easier to discover the global properties of the ionisation. The disadvantage is that pulsars are not observable in other galaxies (except the Magellanic Clouds), so that we cannot use the pulsar dispersion measure to determine the distribution of the electron density, and have to rely on measurements of Hα and other emission lines. As we shall see, it has been possible by these means to observe the DIG in nearby galaxies and to discover that conventional ionisation sources in these galaxies again seem to be inadequate.

6.2 Diffuse Ionisation in nearby Galaxies

A useful source book for this subject is the IAU Symposium No. 144, The Interstellar Disk-Halo Connection in Galaxies (Bloemen 1991), especially the article by Walterbos, and the concluding lecture by Heiles. There is also a good review article by Dettmar (1993) and a recent study by Rand *et al.* (1992). We may summarize their main points as follows:
(i) There is a DIG in some other galaxies, with a temperature $\sim 10^4$ K.
(ii) Its scale height is much greater than that of the HI.
(iii) The emission line ratios [SII]/Hα and [NII]/Hα are higher in the DIG than in HII regions.
(iv) The constancy of the emission line ratios in the DIG rules out shock heating as the main source of ionisation.
(v) Thick Hα disks occur in galaxies which have thick radio continuum disks.
(vi) In a single galaxy the total H α emission is correlated with the total radio emission.
(vii) The Hα in localised regions of a galaxy halo is correlated with the radio emission and the star formation activity in the nearby portion of the disk.

Points (i) to (iv) apply also to our own Galaxy. The remaining points indicate that a key role is played by star formation activity. Walterbos suggests that "galaxies with low star formation activity do not manage to vent material into the halo and thus there are no channels for either the gas or the ionising photons to reach high z distances." A similar remark could be made about regions in a single galaxy where the star formation rate is low. However, we shall see in the next section that in NGC 891 the global star formation rate is probably too low to provide the required ionising photons, so that the role of this activity may be mainly confined to supplying the gas needed to provide the observed free electrons, except perhaps close to the star forming regions themselves. The gas is believed to be blown up through chimneys which may be related to the vertical "worms" in Hα observed above the galactic planes (Norman and Ikeuchi 1989, Koo *et al.* 1992, Rand *et al.* 1992).

6.3 Diffuse Ionisation in NGC 891

NGC 891 is the best-studied nearby galaxy in Hα. It is an edge-on spiral, with an inclination of 88 to 89 degrees. Of all observed galaxies it is the one most like our own (van der Kruit 1984) both in its stellar distribution and its rotation curve. Detailed Hα observations have been made by Rand, Kulkarni and Hester (1990a, b) (hereafter RKH), Dettmar (1990), Keppel *et al.* (1991), and Dahlem, Dettmar and Hummel (1993). The most significant feature of the Hα distribution is its extension away from the plane of the galaxy. This distribution is clearly shown in the Hα picture of NGC 891, taken by RKH, which forms the frontispiece of this book. RKH model this distribution in terms of three components, a "disk", a "halo", and a "bulge", although they stress that this is just a fitting formula rather than a physical decomposition. Each component has the form $< n_e^2 f > e^{-2z/z_0 - 2R/R_0}$, where f is a filling factor. Their halo component has a scale-height z_0 for the n_e distribution (not the n_e^2 distribution, see their erratum in RKH 1990b) of 2.4 kpc (for an assumed distance of 9.5 Mpc) on the north side and 3.2 kpc on the south side.

RKH go on to consider possible ionisation sources for the halo distribution. They examine in particular the possibility that simple cooling of gas accelerated into the halo by chimney action may be responsible (Norman and Ikeuchi 1989). They calculate that the ionising flux associated with the halo Hα at $R = 8.5$ kpc along a column perpendicular to the plane of the Galaxy is 10^6 photons cm^{-2} sec^{-1}. This corresponds to an energy input of at least 2×10^{-5} ergs cm^{-2} sec^{-1} (at ≥ 13.6 eV per ionisation). This is at least a factor 10 to 20 higher than the total energy input by chimneys into the halo of our Galaxy, according to Norman and Ikeuchi. RKH consider that the required rate may be inconsistent with the observed rate of star formation and they suggest that another ionisation agent seems necessary. They add that the extragalactic ionising flux is also too low, and they conclude that a new galactic source is required. Keppel *et al.* (1991) also conclude that the star formation rate is too low to explain the observed ionisation far from the plane.

This conclusion is the same as that of Reynolds and others for our own Galaxy. We shall see in chapter 7 that a similar situation probably exists for the extragalactic ionising flux, which seems to be greater than would be expected from known sources of ionisation. These problems are normally treated independently. In all these cases different additional sources of a conventional kind have been introduced in an ad hoc manner. Although decaying neutrinos are unconventional, this one hypothesis would account naturally for all these ionisation problems, so long as the mass and lifetime of the decaying neutrinos have appropriate values (~ 30 eV and $\sim 2 \times 10^{23}$ sec). This possibility will be discussed in part 3 of this book.

6.4 Ionised Nitrogen in NGC 891

A similar problem is posed by the distribution of ionised nitrogen in NGC 891. This distribution is shown in Fig. 6.1, where the intensity of the emission line [NII] λ6583 is shown (Keppel *et al.* 1991).

It will be seen from this figure that outside the galactic plane the NII closely tracks the HII as measured by the Hα line. In addition, Keppel *et al.* state that the velocities from the [NII] lines agree well with the Hα velocities. This evidence suggests that the nitrogen is ionised wherever the hydrogen is ionised. Since the ionisation potential of nitrogen exceeds that of hydrogen, the same opacity and scale height problems apply as those we discussed in chapter 5. When taken together with the data on ionised nitrogen in our Galaxy, this suggests that photons from decaying neutrinos are ionising nitrogen as well as hydrogen. This possibility will play an important role in part 3 of this book.

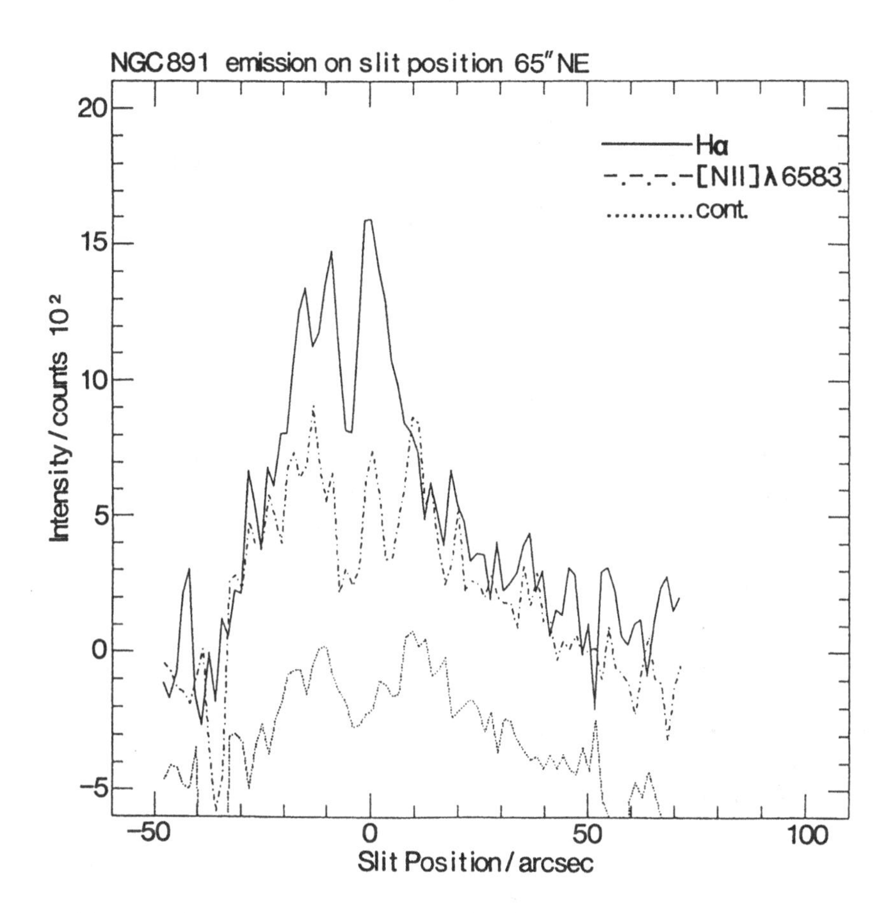

Fig. 6.1 Intensity profiles of Hα, [NII] λ6583 and the continuum perpendicular to the plane of NGC 891 at 65″NE of the nucleus. The continuum is arbitrarily scaled relative to the emission lines. [From Keppel *et al.* 1991].

7
The Intergalactic Flux of Hydrogen-Ionising Photons

7.1 Introduction

The intergalactic flux of hydrogen-ionising photons plays a crucial role in the neutrino decay theory described in part 3 of this book. Accordingly in the present chapter we shall consider various observational estimates of this flux, evaluated at different cosmic epochs. While these estimates are rather uncertain, we shall find that they generally exceed the most recent determinations of the integrated contribution from quasars, which is usually regarded as the main source of the intergalactic ionising flux. Various other photon sources at high red shifts have been proposed to fill the gap, and these are discussed at the end of the chapter. In chapter 11 we shall find that the neutrino decay theory can account for the unexplained flux, but only if various parameters both of the theory and of the universe possess highly constrained values. These values are in general agreement with other estimates of them, and in some cases this agreement is rather precise.

7.2 The Density of Intergalactic Neutral Hydrogen

As soon as quasars of red shift ~ 2 were identified by Schmidt (1965), various authors pointed out that they could be used to probe the density of intergalactic neutral hydrogen. This density could then be used in turn to constrain the intergalactic flux of ionising photons. Consider a layer of neutral hydrogen lying at a red shift z along the line of sight to a quasar of greater red shift Z. Photons emitted by the quasar and reaching the layer with the wavelength of Lyman α would be able to excite neutral hydrogen in the layer to its first excited state. When the excited atom decays to its ground state it would re-emit the Lyman α photon in some general direction. This scattering process would deplete the photon beam propagating towards the observer. Thus there would be an apparent absorption trough in the spectrum of the quasar lying between $1216\mathring{A}$ and $1216(1+Z)\mathring{A}$. For $Z \sim 2$ the long wavelength end of this trough would be accessible to ground-based telescopes, and would lie shortward of the Lyman α emission line produced by the quasar.

The first attempt to detect this absorption trough was made by Gunn and Peterson (1965), and the effect is now named after them. No trough was detected, although the Lyman α excitation process is a resonant one, and so has a large cross-section. Accordingly Gunn and Peterson were able to set a very low upper limit ($\sim 6 \times 10^{-11}$ cm^{-3}) on the density of intergalactic neutral hydrogen at a red shift ~ 2. This density is so low that it is generally assumed that intergalactic hydrogen is highly ionised, rather than almost completely absent.

The expression for the optical depth $\tau_{GP}(z)$ in the absorption trough may be written

$$\tau_{GP} = 4.6 \times 10^5\ \Omega_I y h (1+z)^2 (1+\Omega z)^{-1/2},$$

where Ω_I refers to the total intergalactic density of gaseous hydrogen and y is the neutral fraction of this hydrogen. In particular, if $\Omega = 1$,

$$\tau_{GP} = 4.6 \times 10^5\ \Omega_I y h (1+z)^{3/2} \qquad (\Omega = 1).$$

To determine the value of y we should consider all the intergalactic processes leading to the ionisation of hydrogen (cf. Barcons, Fabian and Rees 1991). We concentrate here on the photoionisation equilibrium in intergalactic space. With the integrated quasar flux in mind, we first assume that the ionisation is entirely due to this flux, which we take to have an intensity $\propto \nu^{-1}$ and a value J at the Lyman limit of $J_{-21} \times 10^{-21}$ ergs cm^{-2} ster^{-1} Hz^{-1}sec^{-1}. To determine the recombination rate α we adopt a gas temperature $\sim 10^4$ K, so that $\alpha \sim 2.6 \times 10^{-13}$ cm^3 sec. The recombination time at $z \sim 2$ then exceeds the Hubble time, but this is not true for significant changes in the low value of n_{HI}. Thus τ_{GP} depends essentially on the contemporary value of J_{-21} rather than on ionisation processes occurring at earlier epochs.

We then obtain

$$y = 7 \times 10^{-7}\ \Omega_I h^2 (1+z)^3 / J_{-21}.$$

Hence

$$\tau_{GP} = 0.3\ \Omega_I^2 h^3 (1+z)^{4.5} / J_{-21} \qquad (\Omega = 1).$$

Note the strong dependence of τ_{GP} on Ω_I, h and particularly z. The value of Ω_I is uncertain. The efficiency of galaxy formation is unknown, so it is better to work directly in terms of Ω_b, which $\sim 0.015\ h^{-2}$. Hence

$$\tau_{GP} = 6.8 \times 10^{-5}\ h^{-1} (1+z)^{4.5} (\Omega_I/\Omega_b)^2 / J_{-21} \qquad (\Omega = 1).$$

We now consider the present observational limits on τ_{GP}. There is an IUE limit at $z = 0$, but in view of the strong dependence of τ_{GP} on z, we shall discuss here only limits for large z. The strongest limits are due to Steidel and Sargent (1987), Jenkins and Ostriker (1991), Webb *et al.* (1992) and Giallongo *et al.* (1992). Following these authors we adopt the limit

$$\tau_{GP} \leq 0.04 \qquad (z \sim 3 \text{ to } 4).$$

At $z \sim 4$ we would have

$$\tau_{GP} = 0.1\ h^{-1} (\Omega_I/\Omega_b)^2 / J_{-21} \qquad (z \sim 4,\ \Omega = 1),$$

and so

$$J_{-21} \geq 2.5 (\Omega_I/\Omega_b)^2 h^{-1} \qquad (z \sim 4,\ \Omega = 1).$$

The lower limit on J_{-21} would be weak if we assumed that most baryons are in galaxies or are not in gaseous form. If, for example, all the dark matter in galaxies were baryonic, one could reduce the limit on J_{-21} by about a factor 10 (Persic and Salucci 1992). Of course at red shifts as high as 4 to 5 the process of galaxy formation may not be complete, so that Ω_I/Ω_b may be relatively large for this reason and even of order unity. In that case, and for $h \sim 0.5$ which would be favoured by consideration of the age of the universe, we would require

$$J_{-21} \geq 5 \qquad (z \sim 4,\ h \sim 0.5,\ \Omega = 1).$$

There should also be a HeI Gunn-Peterson effect. Again no absorption trough is observed; the most sensitive limit so far has been given by Reimers *et al.* (1992). Because the abundance of He is only about one tenth that of H, this test does not constrain J as strongly as does the H test.

7.3 The Ionisation of Lyman α Clouds

The Gunn-Peterson effect discussed above involves Lyman α scattering by neutral hydrogen continuously distributed along the line of sight to a quasar. If one considers instead a discrete cloud of neutral hydrogen on the line of sight, the absorption trough would collapse into an absorption line with a red shift that of the cloud itself. Many such Lyman α absorption lines have been identified in the spectra of quasars, sufficiently many to permit detailed statistical investigations (Sargent *et al.* 1980). We shall now see that the observed properties of these Lyman α clouds enable one to determine a strong lower limit on the intergalactic hydrogen-ionising flux at red shifts $\sim$ 2 to 4. This limit was first derived by Bajtlik, Duncan and Ostriker (1988). They used what is called the inverse or proximity effect, which was discovered by Carswell *et al.* (1982) and was further studied by Murdoch *et al.* (1986) and by many later authors. One first sets up an empirical relation for the number of Lyman α clouds N per unit redshift with a given equivalent width as a function of red shift along a single line of sight. One finds a power law relation of the form

$$dN \propto (1+z)^{\gamma} dz,$$

where $\gamma \sim 1.3$ to 1.9, depending on the equivalent width of the lines (Bechtold 1993). However, as one approaches the quasar the number density of clouds rises less rapidly with red shift than would be expected from this relation. This is the inverse or proximity effect. The origin of this effect is not yet definitely established, although the most likely explanation is that it arises from ionising radiation emitted by the nearby quasar. This explanation is supported by the observed correlation between the absolute luminosity of the quasar and the strength of the proximity effect (Bechtold 1993). In any case, since the quasar's ionising radiation can be measured from the ground for quasars of red shift exceeding 2, one can derive a lower limit on the intergalactic ionising flux incident on the clouds near the quasar, since this is the flux against which the quasar flux is competing to be noticeable. If the proximity effect has a different origin, then the intergalactic ionising flux must be greater.

Bajtlik, Duncan and Ostriker set up a simple photoionisation model in which

$$N_{\rm HI} = \frac{N_0}{1+w},$$

where

$$w = F_Q/4\pi J,$$

and F_Q is the quasar flux at the Lyman limit evaluated at the position of a cloud. On this basis they obtained a lower limit for J_{-21} given by

$$J_{-21} \geq 1 \quad (z \sim 3).$$

Their estimated uncertainty in this lower limit was a factor of 3 either way.

They also found that, within their uncertainty, there was no measureable red shift dependence of J in the range $1.7 < z < 3.8$.

On the basis of a more extended study involving 950 lines, Lu, Wolfe and Turnshek (1991) obtained the same lower limit for J with the same uncertainty. They also stated that "while the exact form of the z dependence of J cannot be determined, the data rule out the models in which log J decreases appreciably below -21.0 at $z > 3$."

A more recent investigation has been made by Bechtold (1993) who allowed for the possibility that the true red shift of the quasar has been underestimated, because of the systematic velocity differences between the lines of high and low ionisation species. She again obtained

$$J_{-21} \geq 1 \quad (z > 2.6).$$

All these estimates of J depend slightly on the spectrum assumed for the intergalactic ionising flux at frequencies above the Lyman limit (Bajtlik, Duncan and Ostriker adopted a $\nu^{-0.5}$ spectrum). What is really determined by the argument is a lower limit on the ionisation rate per hydrogen atom ζ. We give here the value of ζ for the Bechtold result:

$$\zeta \geq 3.7 \times 10^{-12} \text{ sec}^{-1} \qquad (z > 2.6).$$

This limit on ζ will be useful for later comparison with the neutrino decay theory, in which the intergalactic ionising flux has a spectrum very different from a power law.

We should mention that attempts have also been made (Ikeuchi and Turner 1991, Miralda-Escude and Ostriker 1992) to estimate

J_{-21} at $z \sim 0$ from the number of Lyman α clouds observed in the spectrum of 3C 273 by the Hubble Space Telescope (Morris *et al.* 1991, Bahcall *et al.* 1991). These estimates were based on a supposed excess in the number of these clouds which was derived from an older value of 2.3 for the evolutionary parameter γ. With Bechtold's value of 1.3 to 1.9 for γ the discrepancy disappears. We shall therefore not consider these estimates here.

7.4 Sharp Edges of HI Disks in Galaxies

Sensitive 21 cm observations have shown that the HI disks in at least two galaxies (M33 and NGC 3198) have sharp edges (Corbelli, Schneider and Salpeter 1989, van Gorkom *et al.* 1989). In both cases the HI column density N drops from a few times 10^{19} cm^{-2} to a few times 10^{18} cm^{-2} within 1 to 2 kpc, a rate of change much greater than that occurring at higher column densities. The edge of M33 is shown in Fig. 7.1.

Such sharp edges were originally predicted by Sunyaev (1969) and by Longair and Sunyaev (1972) to arise from the ionising effects of the extragalactic photon flux. Detailed calculations of this process were then made by Hill (1974), and by Bochkarev and Sunyaev (1977) and, subsequent to the observational discovery of the sharp edges, by Maloney (1989, 1993) and by Corbelli and Salpeter (1993).

The theory of the effect is simple when the incident flux F is monochromatic and the gas being ionised is smoothly distributed. If its density is n and its thickness is l we would expect $n^2 l$ to decrease with galactocentric distance r. The gas will be mainly neutral at distances for which $\alpha n^2 l > F$, where α is the appropriate recombination co-efficient. It will be mainly ionised at distances for which $\alpha n^2 l < F$. In the region where

$$\alpha n^2 l \sim F$$

the HI column density N will be a steeply decreasing function of r. This edge will occur when

$$N \sim \frac{F}{\alpha n}.$$

If the incident flux has a spectrum extending to energies significantly higher than the ionisation threshold of hydrogen, two com-

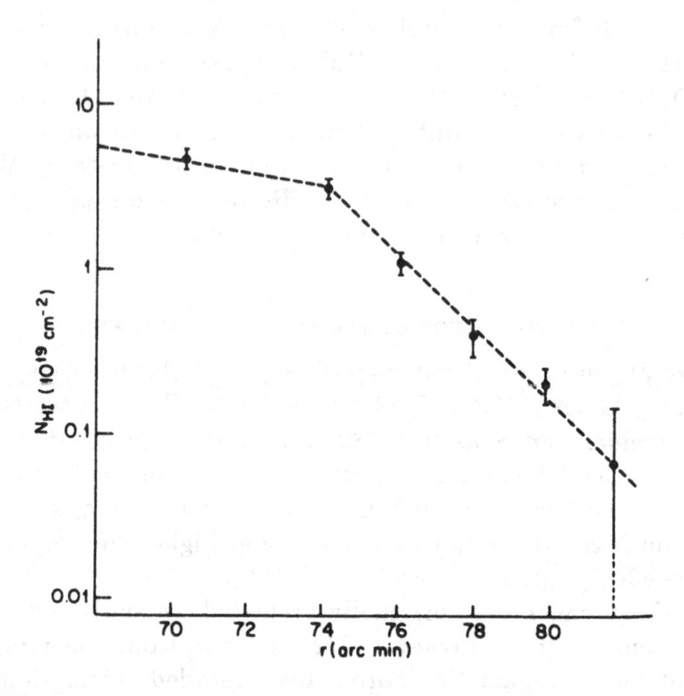

Fig. 7.1 Falloff of the HI in the outer disk of M33. The column density, derived from the HI integrated flux is plotted against the radial distance from the centre of M33 along the northern end of the kinematic major axis. [From Corbelli, Schneider and Salpeter 1989: Reproduced courtesy of *The Astronomical Journal*].

plications arise. The first is a result of the rapid decrease of the ionisation cross-section with photon energy ($\sim E^{-3}$). The higher energy photons can then penetrate further into the HI layer and tend to smooth out the sharp HI/HII transition which would be produced by a monochromatic ionising flux.

The second complication arises because helium in the gas layer will be ionised by sufficiently high energy photons. When the helium recombines some of the photons produced are energetic enough to ionise hydrogen. These and other similar complications have been included in the detailed model calculations of Corbelli and Salpeter (1993) and Maloney (1993).

There is, however, another complication which is much harder to quantify. As pointed out by Longair and Sunyaev (1972), the simple theory which we have outlined would break down if the distribution of HI is strongly clumped. If n fluctuates violently, regions where n is much greater than its mean value would contribute a column density of HI in the "forbidden range", and the sharp edge would be smeared out. Longair and Sunyaev and also Maloney argued that such clumping is unlikely to occur in the outer regions of galaxies. However, Henderson, Jackson and Kerr (1982) and Merrifield (1992) have stressed that in the outer regions of our own Galaxy the distribution of HI does tend to be strongly clumped.

Yet another complication could arise from the cloudy nature of the medium being ionised. It would be relatively easy for a set of, possibly transient, clouds lying outside the main HI layer to absorb a significant fraction of the incident ionising flux, rendering it irrelevant to the existence of sharp HI edges. Consider, for instance, a set of clouds with, say, a density within each cloud $\sim 1\ \mathrm{cm}^{-3}$, a size $\sim 10^{17}$ cm, a temperature ~ 100 K, and with $\sim$ one cloud per cubic pc. There would then be ~ 10 such clouds along a line of sight through a cloud layer of total thickness ~ 6 kpc, which would be enough to absorb most of the ionising flux associated with the integrated quasar radiation background. The total column density in these clouds would only be $\sim 10^{18}\ \mathrm{cm}^{-2}$, which is less than the smallest reliable positive detection of an HI column density in the existing observations on sharp edges. These clouds would also be unobservable in absorption at 21 cm. Admittedly there is some fine tuning in the postulated properties of these clouds, but it does seem desirable to extend the models to include a reasonable degree of cloudiness.

It might, of course, be argued that the observed existence of sharp edges is itself evidence that the ionised medium is sufficiently smoothly distributed in the outer regions of M33 and NGC 3198. This might be so if the only possible ionising agency were the incident external photon flux. However, we shall see that, in our decaying neutrino theory, an alternative internal ionising flux is available. Ironically, in this case the predicted edge would be sharper than the observed one unless the medium is sufficiently cloudy.

For completeness we record here the calculated incident ionising flux for a smooth medium. To obtain an edge setting in at 3×10^{19} cm^{-2} we would require

$$J_{-21} \sim 4\times10^{-2} \text{ to } 4\times10^{-1} \qquad (z \sim 0),$$

or, equivalently

$$F \sim 10^4 \text{ to } 10^5 \text{ photons cm}^{-2} \text{ sec}^{-1} \qquad (z \sim 0).$$

A full treatment of the problem would involve also the steepness of the edge, which has been discussed by Corbelli and Salpeter and by Maloney. We cannot go into this question here.

7.5 Hα Measurements

A direct method of estimating the intergalactic hydrogen–ionising flux F is to measure the Hα surface brightness of clouds opaque to F exposed to this flux. Each hydrogen recombination following an ionisation produces on average 0.46 Hα photons (Martin 1988, case B, $T \sim 10^4$ K). Thus the ionisation rate can be determined from $H\alpha$ measurements.

In practice this method suffers from a number of uncertainties. On the one hand there may be significant ionisation sources within the cloud being observed or in a surrounding galaxy. On the other hand the cloud may lie in a shadow (Kutyrev and Reynolds 1989). This latter point is particularly serious, since as we saw in the previous section, it is relatively easy for a set of clouds to absorb ionising radiation whose energy is close to the Lyman limit. For these reasons, it is not surprising that the literature contains a number of conflicting results for the intergalactic flux, which range over a factor of 10. In this book we shall take the point of view that, unless there is reason to believe that internal sources are important, the larger values are the more relevant ones because they are presumably less subject to shadowing. In order of increasing flux limits the published results are as follows:

1) Songaila, Bryant and Cowie (1989) observed Hα in a high velocity HI cloud in the halo of the Milky Way (cloud complex C). Some of the implied ionising flux may be produced in the galactic halo itself. They obtained an upper limit on the extragalactic ionising flux F of

$$F \leq 6 \times 10^4 \text{ cm}^{-2} \text{ sec}^{-1} \qquad (z = 0).$$

2) Stocke *et al.* (1991) observed an HI gas cloud lying between the galaxy NGC 3067 and the quasar 3C232. They failed to detect any Hα, which translates into an upper limit on F of

$$F < 8 \times 10^4 \text{ cm}^{-2} \text{ sec}^{-1} \qquad (z \sim 0).$$

3) Kutyrev and Reynolds (1989) observed a very high velocity cloud in Cetus and found weak emission at Hα. Their upper limit on F was

$$F \leq 2 \times 10^5 \text{ cm}^{-2} \text{ sec}^{-1} \qquad (z = 0).$$

4) Reynolds *et al.* (1986) obtained a 4σ detection of Hα from an intergalactic HI cloud in Leo. Their upper limit on F was

$$F \leq 6 \times 10^5 \text{ cm}^{-2} \text{ sec}^{-1} \qquad (z \sim 0).$$

5) Münch and Pitz (1990) observed Hα lines in two high galactic latitude clouds known to emit 21 cm lines. Their resulting value for F (including radiation from the galactic halo itself) was

$$F \sim 8.5 \times 10^5 \text{ cm}^{-2} \text{ sec}^{-1} \qquad (z = 0).$$

6) Reynolds (1984) derived from his Hα measurements on diffuse gas in the Milky Way a total ionising flux along a line of sight through the sun at right angles to the galactic plane of

$$\sim 4 \times 10^6 \text{ cm}^{-2} \text{ sec}^{-1}.$$

Since this result and that of Munch and Pitz includes the ionising flux emitted in the galactic halo and disk, we prefer to adopt for the extragalactic ionising flux the value derived from the Hα observation by Reynolds *et al.* of the intergalactic HI cloud in Leo. We shall therefore take for this ionising flux F_{ext} an upper limit

$$F_{ext} \leq 6 \times 10^5 \text{ cm}^{-2} \text{ sec}^{-1} \qquad (z \sim 0).$$

This limit will play a crucial role in the neutrino decay theory (chapter 11).

7.6 Lyman α Measurements

Another direct method of estimating F is to measure the Lyman α emission induced by it (Hogan and Weymann 1987). If the system concerned has an HI column density exceeding 10^{17} cm^{-2} (for example, a damped Lyman α system in a quasar absorption spectrum (Wolfe *et al.* 1986)) most of the incident ionising photons produce Lyman α photons after they have traversed a column density of order 10^{17} cm^{-2}. The interior HI then scatters nearly all the Lyman α photons back into intergalactic space (Neufeld 1990). The ionised outer layers of the damped Lyman α systems have such small column densities that absorption by dust can be neglected (Charlot and Fall 1991).

The theory connecting the induced surface brightness S_α in Lyman α to the incident ionising flux J_ν has been given by Songaila, Cowie and Lilley (1990), Charlot and Fall (1991), Wolfe *et al.* (1992) and Binette *et al.* (1993). They find that

$$S_\alpha = 0.6\nu_L J/(1+z)^4,$$

where $J_\nu \sim \nu^{-1}$, and z is the red shift of the emitting system. If there is reason to believe that stars in the emitting system are contributing significantly to J, this method would provide an upper limit to F rather than an actual measurement of it.

Until recently attempts to detect Lyman α emission from damped Lyman α systems have been unsuccessful (e.g. Smith *et al.* 1989, Deharveng *et al.* 1990, Lowenthal *et al.* 1990, 1991, Pettini *et al.* 1990). Hunstead *et al.* (1990) claimed to have observed Lyman α emission centred in the black core of a Lyman α absorption line at $z = 2.4654$ in the spectrum of the quasar H 0836+113. This claim has been contested by Wolfe *et al.* (1992), who, however, apparently detected Lyman α emission from an extended object detected within 4″ of the quasar. They obtained

$$S_\alpha = 2.6 \times 10^{-18} \text{ ergs cm}^{-2} \text{ sec}^{-1} \text{ arcsec}^{-2}$$

which implies the following upper limit on the extragalactic ionising flux J_{-21} (Wolfe *et al.* 1992, Binette *et al.* 1993):

$$J_{-21} \leq 11 \qquad (z = 2.5).$$

This upper limit is about ten times greater than the central value of the lower limit derived by Bajtlik, Duncan and Ostriker (1988) from the proximity effect in Lyman α clouds. It may be possible in future to distinguish spectroscopically between Lyman α emission lines excited by the intergalactic ionising flux and by starburst photons emitted deep within the damped Lyman α systems (Wolfe *et al.* 1992, Binette *et al.* 1993). In both cases the emergent spectrum of the Lyman α line would consist of two humps symmetrically placed about the line centre, but the centroids of the humps would be more widely separated in the latter case.

Table 7.1.

z	0	2	4
ζ $(10^{-12}\ \text{sec}^{-1})$	≤ 3.6	≥ 3.7 ≤ 40	≥ 17

7.7 Summary of Observational Estimates of the Intergalactic Ionising Rate

To facilitate comparison with various theoretical estimates of ζ, we summarize in table 7.1. the observational results which we have discussed. The estimate for $z \sim 4$ is based on the Gunn-Peterson effect, and assumes that at that red shift $\Omega_I \sim \Omega_b$ and that $h \sim 0.5$ and $\Omega \sim 1$.

7.8 The Extragalactic Far Ultraviolet Background

Since the interstellar medium is opaque to the extragalactic hydrogen-ionising flux down to a wavelength $\sim 200\mathring{A}$ one cannot hope to detect this flux directly at the Earth. It is therefore helpful to supplement the indirect studies of the flux, described in the earlier sections of this chapter, with direct measurements of the extragalactic background at nearby longer wavelengths, say 912 to $2000\mathring{A}$, which would be expected to penetrate the Galaxy despite appreciable attenuation due to interstellar dust. The flux F_u expected at these wavelengths and having the same origin as the ionising flux F depends, of course, on its spectrum, which is unknown. To see what is involved we may make a rough estimate as follows. The usual units in which the ultra-violet continuum is measured are photons $\text{cm}^{-2}\ \text{sec}^{-1}\ \text{ster}^{-1}\text{Å}^{-1}$ (Continuum Units or CU). If we adopt the value $6 \times 10^5\ \text{cm}^{-2}\ \text{sec}^{-1}$ for F_{ext} as discussed earlier, and assume a smoothly varying spectrum out to $2000\mathring{A}$, we could consider a tentative effective bandwidth in this wavelength region of, say, $300\mathring{A}$. We would then obtain the rough estimate

$$F_u \sim 200\ \text{CU},$$

if galactic absorption by dust is neglected.

Of course, this estimate is uncertain by a factor of at least 3 either way if our choice of F_{ext} is correct, but it provides a useful starting-point for comparison with the attempts which have been made to detect an extragalactic far uv background.

Unfortunately these attempts are rather controversial. Apart from the difficulties of the measurements themselves, which have to be made from space, there are large but uncertain corrections for the scattering by dust of radiation produced by stars in the Galaxy. One must also allow for two-photon emission from the Reynolds layer, and for far uv emission from galaxies whose ionising emission is absorbed within each galaxy. Finally one must correct for the attenuation of F_u by interstellar dust.

Fortunately for us, the extensive literature on these questions has recently been reviewed in detail by Bowyer (1991), Henry (1991) and Wright (1992). We may summarise their discussion by saying that no extragalactic flux has been confidently established at far uv wavelengths, with the possible exception of radiation from spiral galaxies which was identified by its angular correlation properties (Martin and Bowyer 1989) and, since these reviews were written, by direct galaxy counts at 2000Å (Milliard et al 1992). The resulting upper limit on F_u is still controversial, but is spanned by the range

$$F_u < 100 \text{ to } 300 \text{ CU}.$$

We should also mention the recent work of Onaka and Kodaira (1991) and of Murthy, Henry and Holberg (1991). The first authors found that

$$F_u > 50 \text{ CU},$$

when allowance is made for the contribution from spiral galaxies. The second authors found no evidence for the presence of a diffuse radiation field between 912 and 1100Å at a level of 100 to 200 CU. Since these latter observations were carried out at galactic latitudes ranging from 11.5 to 53.7 degrees, it is necessary to correct for the absorption of F_u by interstellar dust.

This correction is particularly important at far uv wavelengths where the dust extinction is very high. Following Draine and Lee (1984) and Desert, Boulanger and Puget (1990) we adopt an extinction cross-section at 2000Å of 2.4×10^{-21} cm^2 per H atom. We then derive an optical depth at high galactic latitudes of ~ 0.65, and at intermediate latitudes of ~ 1, corresponding to an attenuation varying from ~ 2 to ~ 3. When allowance is made for this attenuation, there is no discrepancy between our rough estimate of F_u and the existing observations. A relatively small improvement in the observations would, however, either lend support to our estimate of F_u, and so perhaps of F_{ext}, or cast considerable doubt on them.

7.9 The Intergalactic Flux of Ionising Photons from Quasars

In this section we examine whether quasars can account for the intergalactic ionising flux which we have estimated above. It was first suggested by Arons and McCray (1970) that the absence of a Gunn-Peterson effect in the spectra of quasars is due to the integrated radiation from the quasars themselves, and this possibility has been much discussed since then. Two questions are involved here. The first question is whether the HII regions surrounding the individual quasars overlapped at sufficiently early epochs (Arons and Wingert 1972, Shapiro and Giroux 1987, Donahue and Shull 1987, Madau and Meiksin 1991). This question is particularly pressing now, when we know that no absorption trough has been observed out to a red shift as large as 4.9.

The second question arises if we accept that the overlap did occur at suitably large red shifts. Is the integrated ionising flux from quasars large enough to account for the ionisation of both the intergalactic medium and Lyman α clouds? The answer to this question depends critically on what is assumed for the cosmic evolution of the quasar luminosity function. The existence of such evolution was first pointed out by Sciama and Rees (1966), and it was then confirmed by Schmidt (1966) on the basis of more extensive data. There is a rapid evolution out to $z \sim 2$, but for greater red shifts the amount of evolution has been the subject of recent controversy.

A careful analysis of the problem has been made by Madau (1992) who has used the most recent surveys of very high red shift quasars, and has also evaluated the absorption occurring at high red shifts from the presence of Lyman α clouds and Lyman limit systems. Madau's results are shown as a function of red shift

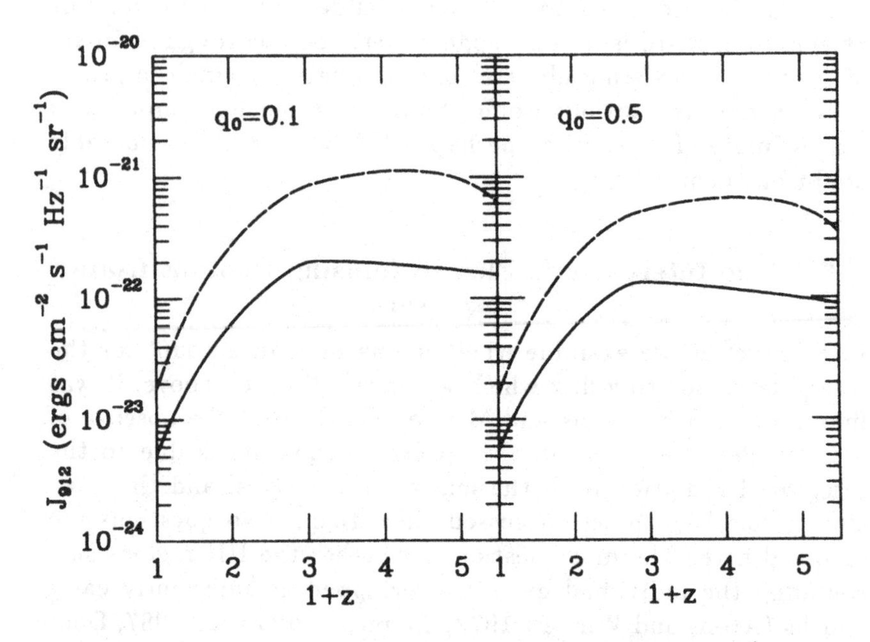

Fig. 7.2 Quasar contribution to the intergalactic flux at 912Å as a function of redshift for $q_0 = 0.1$ and 0.5. The solid lines allow for absorption by Lyman α clouds and Lyman limit systems. [From Madau 1992].

in Fig. 7.2.

The implied values of ζ are significantly less than the observational estimates given in table 1 for $z \sim 2$ and 4†, and also significantly less than the value required to account for sharp HI edges in galaxies at $z \sim 0$ in the theory assuming a smooth distribution of the ionised gas.

There remains the possibility that a large number of individually faint quasars may collectively contribute a significant fraction of J. Nevertheless it is widely believed that other ionising sources may be needed to account for the observed value of J. This possibility has been recognised for some time, and a number of more

† For a recent contrary view see Meiksin and Madau (1993). However, this discussion predates the important work of Bechtold (1993).

or less conventional alternative sources have been proposed. Some of these are briefly described in the next section.

7.10 Alternative Sources of Intergalactic Ionising Photons

A number of suggestions have been made for possible alternative sources of intergalactic ionising photons. These sources are usually placed at red shifts exceeding 5, so that there is not much observational control over possible models. One exception to this statement concerns massive stars, whose contribution to the metallicity of their surroundings can be related to their output of ultra-violet radiation (e.g. Silk, Wyse and Shields 1987, Shapiro and Giroux 1989, Songaila, Cowie and Lilly 1990, Miralda-Escude and Ostriker 1990). Proposed sources include population III stars (Carr, Bond and Arnett 1984), elliptical galaxies (Bechtold *et al.* 1987), dwarf galaxies (Silk, Wyse and Shields 1987), starbursting galaxies (Songaila, Cowie and Lilly 1990), quasars concealed by dust obscuration in intervening galaxies (Miralda-Escude and Ostriker 1990), reflection-dominated hard x-ray sources which may also be responsible for the x-ray background (Fabian *et al.* 1990), accreting black holes (Collin-Souffrin 1991) and supernova-driven winds from early galaxies (Tegmark, Silk and Evrard 1993).

A related issue is whether these sources can produce enough hard radiation to account for the high ionisation (e.g. in the form of SiIV, CIV and NV) found in absorbing clouds in quasar spectra. According to Steidel and Sargent (1989) this represents a problem for some of the proposed sources. However, Madau (1991) claims that this is no longer so when allowance is made for the effect on the integrated spectrum of absorption produced by intervening Lyman α clouds. In any case, it is possible that the high ionisation needing explanation may result not from a radiation field, but from collisional processes in the hot gas associated with "galactic fountains" in the absorbing clouds (Martin and Bowyer 1990, Sembach, Savage and Massa 1991).

The role of the intergalactic ionising flux is characteristically different in the decaying neutrino theory. If decaying galactic neutrinos are indeed mainly responsible for the diffuse ionisation of the Milky Way, then decaying cosmological neutrinos can readily produce an intergalactic ionising flux of the needed intensity both at $z = 0$ and at $z \sim 2$ to 5. In fact the requirement that they do not overproduce ionising photons will lead to stringent constraints on the energy of the decay photons, on the rest mass of the decaying neutrinos, and on the Hubble constant, which have a precision of about 1%.

Part Three

Neutrino Decay and Ionisation in the Universe

8
The Radiative Decay of Massive Neutrinos

8.1 Introduction

If the different neutrino types have different rest masses we would expect a more massive neutrino ν_1 to decay into a photon and a less massive neutrino ν_2:

$$\nu_1 \longrightarrow \gamma + \nu_2.$$

We propose using these decay photons to solve all the ionisation problems described in part 2 of this book. The decay is achieved by ν_1 first virtually transforming itself into two charged particles (for example, a W boson and a charged lepton). One of these charged particles can then emit a photon, after which the two charged particles recombine into a ν_2. The kinematics of this reaction are simple. The photon produced is monochromatic, and its energy E_γ in the rest frame of ν_1 is determined by conservation of energy and momentum in the decay to be given by:

$$E_\gamma = \frac{1}{2} m_1 \left(1 - \frac{m_2^2}{m_1^2}\right).$$

If $m_2 << m_1$, which we shall see is rather likely (page 144), we would have the simple relation

$$E_\gamma \sim \frac{1}{2} m_1,$$

each "zero rest mass" particle in the decay carrying off half the available energy in the rest frame of ν_1. For the moment we shall assume that this relation holds good. Later on we will find observational evidence that $m_2/m_1 \lesssim 0.2$ in our neutrino decay theory.

8.2 Decaying Neutrinos in Astronomy and Cosmology

Whether the radiative decay of neutrinos is of interest in astronomy or cosmology depends, of course, on the lifetime τ. It was first pointed out by Cowsik (1977) that lines of sight through the cosmological sea of neutrinos are so great ($\sim 10^{28}$ cm) that even an apparently large value of τ might lead to an observable flux of decay photons. Cowsik also pointed out that if m_1 were of the general order of 1 eV then the decay photons would lie in the optical range.

This idea was extended by de Rujula and Glashow (1980) to the more cosmologically interesting range for m_1 of 10 to 100 eV, in which case the decay photons would have energies in the range 5 to 50 eV and so would lie in the ultra-violet. They also made theoretical estimates of the lifetime τ on the basis of various particle physics models of the decay. Such estimates had previously been made quite independently of astronomical considerations, and there is now a large literature on this subject which we will consider shortly.

One can place a lower limit on τ by requiring that the high energy neutrinos emitted by white dwarfs and supernovae, and especially supernova 1987A, do not produce x or γ rays in excess of observed limits. In these calculations one must allow for the large effect of time dilation on the appropriate value of τ. One finds in this way that

$$\frac{\tau}{m_\nu} \gtrsim 10^{16} \text{ to } 10^{18} \text{ sec/ eV}$$

(Cowsik 1977, von Feilitzsch and Oberauer 1988, Chupp *et al.* 1989, Kolb and Turner 1989).

It is also possible to place a stronger lower limit on τ from the cosmological distribution of neutrinos, since the observed visible and ultra-violet backgrounds must not be exceeded by the flux of decay photons. In calculating this flux both Stecker (1980) and Kimble, Bowyer and Jakobsen (1981) pointed out that the integrated flux I_α of decay photons from cosmological neutrinos would be drawn out by the red shift into a continuous spectrum. They obtained (in the absence of absorption)

$$I_{\lambda>\lambda_0} = \frac{n_\nu(z=0)}{\tau}\frac{c}{H_0}\frac{\lambda_0^{3/2}}{\lambda^{5/2}}\left[1+(2q_0-1)\left(1-\frac{\lambda_0}{\lambda}\right)\right]^{-1/2},$$

where λ_0 is the rest wavelength of the decay photons, and λ is the observed wavelength. Even if the emitted photons have $\lambda < 912\mathring{A}$ ($E_\gamma > 13.6$ eV), so that they can ionise hydrogen and therefore be unable to reach us by penetrating the neutral hydrogen in the Galaxy, the red shifted tail of the spectrum would reach to $\lambda > 912\AA$. Using this fact, and observed values for the ultra-violet background which were available at the time, they were able to derive a lower limit for τ:

$$\tau > 10^{22} \text{ to } 10^{23} \text{ secs} \quad (E_\gamma \sim 5 \text{ to } 50 \text{ eV}).$$

More recent observational estimates of the ultra-violet background have been discussed on page 112. After allowance for absorption by interstellar dust, these estimates lead to the slighter stronger limit (Sciama 1991b, Overduin, Wesson and Bowyer 1993, Dodelson and Jubas 1993)

$$\tau \geq 2 \times 10^{23} \text{ secs} \quad (E_\gamma \sim 15 \text{ eV}).$$

It is a striking fact that an astronomically interesting value of τ, say in the range 10^{23} to 10^{24} secs, is about a million times longer than the age of the universe–ample testimony to Cowsik's point that the line of sight being used is a very long one. Such a large value would also mean that at the present time most of the neutrinos in the universe would not yet have decayed. There are in fact many schemes in the literature in which particles produced in the hot big bang decay on a timescale short compared with the age of the universe. We shall not consider such schemes in this book.

If $E_\gamma > 13.6$ eV one can also limit τ by demanding that the decay photons should not produce excessive hydrogen ionisation. Such a limit was obtained by Melott and Sciama (1981) who used the existence of the high-velocity clouds of neutral hydrogen which are observed in the halo of our Galaxy. Although the cosmologically produced ionising photons might be absorbed before reaching the clouds, the dark halo of the Galaxy could itself consist mainly of massive neutrinos. Their decay would then lead to the limit

$$\tau \geq 10^{23}\left(\frac{T}{10^4}\right)^{3/2}\left(\frac{0.6 \text{ cm}^{-3}}{\mathrm{n_e}}\right)^2\left(\frac{1 \text{ kpc}}{\mathrm{d}}\right)\left(\frac{0.05 \text{ rad}}{\varphi}\right)\left(\frac{30 \text{ eV}}{\mathrm{m}_\nu}\right)N$$

where τ is in seconds, T is the cloud temperature, n_e its electron density, d its distance, φ its angular radius and N the average number of ionisations caused by a given photon. Again it would seem that the astronomically interesting range of τ is 10^{23} to 10^{24} secs.

Subsequently Sciama and Melott (1982) pointed out that ionisation effects in the interstellar medium produced by decaying Galactic dark matter particles (neutrinos or photinos) might actually be observable, and this idea was further pursued in a number of papers (Sciama 1982a, b, d, 1984). All this work concerned the relatively rare ions CIV and SiIV; to produce the former would require $E_\gamma > 47.9$ eV. It was also suggested (Rephaeli and Szalay 1981, Sciama 1982c, Salati and Wallet 1984, Melott 1984, Melott, McKay and Ralston 1988) that the ionisation of hydrogen and helium in the intergalactic medium may be mainly due to the decay of cosmologically distributed dark matter. The anomalously large ionisation of Lyman α clouds at red shifts 2 to 4 (Bajtlik, Duncan and Ostriker 1988) was also attributed to decay photons from cosmological dark matter (Sciama 1988).

The work of this period which comes closest to the theory which we are about to describe is that by Melott, McKay and Ralston (1988), who pointed out that a variety of phenomena in astronomy and cosmology could be explained if the decaying neutrinos have a mass ~ 30 eV and a lifetime $\sim 10^{24}$ sec. The present author then took the further step (Sciama 1990a) of suggesting that the widespread ionisation observed in the Milky Way, and described in chapter 5, is mainly due to decaying neutrinos which provide most of the dark matter in our Galaxy. As we shall see in the next chapter, this suggestion would require the neutrinos to have the rather shorter decay lifetime of 2 to 3×10^{23} sec, which is close to the lowest permitted value. Although this step may seem to involve a relatively small reduction in lifetime, it turns out that such a reduction would have major consequences. It would lead to a very closely specified theory with a strong impact on many phenomena of astronomy and cosmology. We devote the rest of this book to a study of this theory, beginning in the next section with the description of a particle physics model which would lead to the required radiative lifetime of 2 to 3×10^{23} secs for a 30 eV neutrino.

8.3 Theoretical Estimates of the Radiative Lifetime of Neutrinos

The first reference to the radiative decay of massive neutrinos was made by Nakagawa *et al.* (1963), who estimated τ for a neutrino of mass 1 MeV. Subsequently many calculations were made for both Dirac and Majorana neutrinos. The early calculations were carried out without using gauge theory, and some of them contain serious errors. The first correct calculations using the gauge theory of the Salam-Weinberg standard model were made by Marciano and Sanda (1977) and by Petcov (1977a, b). This work has been reviewed and extended by Pal and Wolfenstein (1982).

It turns out that in this model two effects exist which suppress the decay to such an extent that it would become astronomically uninteresting. The first effect is called GIM suppression (GIM stands for Glashow, Iliopoulos and Maiani 1970) while the second is the helicity flip which occurs on the external ν line in the appropriate Feynman diagrams. As a result one obtains a lifetime $\tau \sim m_1^{-5}$, and for $m_1 \sim 30$ eV one has

$$\tau \sim 10^{30} \text{ secs},$$

which is a million times too long to be astronomically interesting.

Despite its great success, the standard model is not regarded as the last word by particle physicists. In particular, it contains a large number of undetermined and independent parameters. This number can be substantially reduced in various extensions of the standard model involving grand unification, supersymmetry and other generalisations. It is of interest to see whether any of these models lead naturally to the astronomically interesting lifetime of 10^{23} to 10^{24} secs for a 30 eV neutrino. One could even take the point of view, as I do, that the astronomical and cosmological evidence for $\tau \sim 2$ to 3×10^{23} secs is strong enough to be used as a clue towards the choice of the correct particle theory.

To see what is involved here it is helpful to begin by considering a general relation which holds between the decay lifetime τ and another manifestation of an electromagnetic connexion between two neutrino flavours, namely, the so-called transition magnetic moment μ_{12}, which also has direct astronomical significance. To understand the meaning of this quantity it is helpful to recall that when a particle has an ordinary magnetic moment, its spin will precess in the presence of a magnetic field B. If the particle ν_1 possesses a transition magnetic moment, then while it precesses it picks up a component of $\bar{\nu}_2$, at a rate depending on $(\mu_{12}B)^{-1}$ and the rate of variation of B (Schechter and Valle 1981). This process, called flavour changing spin precession, is reminiscent of oscillations between two neutrino flavours, and has a similar formal structure. In particular it can be enhanced by the presence of matter (Akhmedov 1988, Lim and Marciano 1988) in a manner analogous to the MSW effect, and this has been proposed as an alternative solution of the solar neutrino problem (Akhmedov, Lanza and Petcov 1993).

There is a simple relation between τ and μ_{12} which holds quite generally, and does not depend on the particular theory used to calculate them (Marciano and Sanda 1977). This relation is

$$\tau = \frac{8\pi}{\mu_{12}^2} \frac{1}{m_1^3} \left[1 - \left(\frac{m_2}{m_1}\right)^2\right]^{-3}.$$

In terms of the Bohr magneton $\mu_B = (eh/2m_e c)$ one can write this relation (assuming $m_2/m_1 \ll 1$)

$$\tau = 10^{23} \left(\frac{30eV}{m_1}\right)^3 \left(\frac{10^{-14}\mu_B}{\mu_{12}}\right)^2 \text{ sec.}$$

Thus a lifetime $\tau \sim 10^{23}$ sec would be associated with a transition moment $\mu_{12} \sim 10^{-14}\mu_B$ if $m_1 \sim 30$ eV.

This is astronomically interesting because one can constrain μ_{12} by demanding that plasmons in a red giant should not produce $\nu_1\nu_2$ pairs so rapidly near the helium flash that the escape of these neutrinos would cool the star unacceptably fast. This, and other similar constraints, have been discussed by Raffelt (1990b), Raffelt and Weiss (1992) and Castellani and Degl'Innocenti (1993). They find that

$$\mu_{12} \lesssim 10^{-12}\mu_B.$$

Raffelt also pointed out that this leads to the following lower limit on τ

$$\tau > 10^{19} \left(\frac{30 \text{ eV}}{m_1}\right)^3 \text{ sec.}$$

(cf. Nussinov and Rephaeli 1987, Ralston, McKay and Melott 1988).

While this lower limit on τ is safely 10^4 times less than the value which we are adopting, we note that in terms of the transition moment μ_{12} we are only a factor 100 away from a value which would be important for stellar evolution. It would be interesting if this gap could be bridged in some special stellar systems containing a substantial magnetic field, such as the sun (Akhmedov, Lanza and Petcov 1993) or, more plausibly, supernovae.

A further upper limit on μ_{12} has recently been obtained by an interesting, but rather model-dependent, argument due to Enqvist *et al.* (1992). They point out that a primordial magnetic field would cause left-handed Dirac neutrinos to oscillate into right-handed neutrinos which could then contribute an unacceptable extra neutrino type at the epoch of primordial nucleosynthesis. One can derive in this way an upper limit on both the magnetic moment and the transition moment of neutrinos in terms of the strength of the seed field. Enqvist *et al.* derive for a matter-enhanced resonant conversion (Akhmedov 1988, Lim and Marciano 1988) the upper limit:

$$\mu_{12} \leq 5 \times 10^{-15} \left(\frac{1 \text{ eV}^2}{\Delta m^2} \right)^{5/12} \frac{10^{12}}{C} \mu_B,$$

where C is a phenomenological parameter to be fitted in such a way that typical galactic magnetic fields will be generated from the seed field by dynamo action. They point out that usually C is chosen to be $\sim 3.7 \times 10^9$, but that there exists evidence that it may be considerably larger. Provisionally adopting this value, and taking $\Delta m^2 \sim m_1^2$ we would obtain

$$\mu_{12} \leq 5 \times 10^{-14} \left(\frac{30 \text{ eV}}{m_1} \right)^{5/6} \mu_B$$

and so

$$\tau \geq 2.5 \times 10^{22} \left(\frac{30 \text{ eV}}{m_1} \right)^{4/3} \text{ secs.}$$

This lower limit on τ is tantalisingly close to our previous lower limit based on the observed ultra-violet background. However, it should be remembered that the origin of galactic magnetic fields is still not understood, and may not even involve the dynamo amplification of a primordial field.

We now consider extensions of the standard model which could lead to $\tau \sim 2$ to 3×10^{23} secs for $m_1 \sim 30$ eV. A number of possibilities have been discussed (e.g. Maalampi and Roos 1991). However, we will concentrate here on two models which are closely similar to one another but were independently and simultaneously devised (partly at the same institute, SISSA!) (Roulet and Tommasini 1991, Gabbiani, Masiero and Sciama 1991). Both these models are supersymmetric. Supersymmetry (SUSY) is a theory which contains symmetry operations between bosons and fermions (for a review of this theory with extensive references see Nilles 1984). The existence of these symmetry operations leads to the introduction of new particles (or sparticles) each of which is the supersymmetric partner of a known particle and has a spin differing by half a unit. For example, the partner of the photon (spin 1) is the photino with spin $\frac{1}{2}$, while the selectron has spin zero. Since a spin zero particle with the same mass as the electron would have been easily detected, one deduces that SUSY must be broken at low energies, in a manner familiar from modern gauge theories of elementary particles. This breaking of SUSY would lead to the photino having a non-zero rest mass, which permits it to be a candidate for the dark matter in astronomy and cosmology.

The existence of this new set of particles in SUSY theories has led to the introduction of a new quantum number which can be used to classify particles. This new quantum number is called R parity and is defined by

$$R = (-1)^{2S+3B+L},$$

where S is the spin, B the baryon number and L the lepton number of each particle. Known particles in the standard model all have $R = +1$, while their SUSY partners all have $R = -1$. In the early versions of SUSY it was assumed that R parity is multiplicatively conserved by all interactions. If this were the case, SUSY particles could only be produced in pairs, and the lightest SUSY particle would be absolutely stable. In addition lepton number would be conserved.

It turns out that if R parity is conserved one could only achieve $\tau \sim 2$ to 3×10^{23} sec if at the same time the τ meson would decay into $e + \gamma$ and $\mu + \gamma$ at a faster rate than is permitted by the present experimental bounds, in spite of the looseness of such bounds (Hayes *et al.* 1982, Keh *et al.* 1988). However, couplings which break R parity were introduced into SUSY theory in 1983, and there is now a large literature on this subject (some references to which are given by Roulet and Tommasini). These authors studied the so-called Minimal Supersymmetric Model with broken R parity and showed that in this theory one can obtain a decay lifetime $\tau \sim 2$ to 3×10^{23} sec for $m_1 \sim 30$ eV with either ν_τ or ν_μ playing the part of ν_1.

Gabbiani, Masiero and Sciama proceeded slightly differently by carrying out their analysis in the framework of spontaneously broken $N = 1$ supergravity models (Nilles 1984) for both broken and unbroken R parity. They again found that in theories with broken R parity they could achieve $\tau \sim 2$ to 3×10^{23} sec for $m_1 \sim 30$ eV without violating any known bounds.

One difficulty with theories of this type, however, is that they lead to neutrino oscillations which are barely consistent with the experimental bounds (K. Enqvist, A. Masiero and A. Riotto 1992). In a recent extension of these theories Tommasini (1992) has shown that the Barr-Freire-Zee (1990) mechanism would suppress these oscillations and would automatically associate with the transition moment μ_{12} a mass scale for the lighter ν_2. For $\mu_{12} \sim 10^{-14}\mu_B$ this mass scale is naturally close to the value $\sim 10^{-3}$ eV associated with the MSW solution of the solar neutrino problem (page 144).

One potential difficulty with theories of this type concerns their implications for our understanding of baryosynthesis (Campbell *et al.* 1991, 1992, Fischler *et al.* 1991). These difficulties could be resolved if the epoch of baryosynthesis were delayed until the electroweak transition took place. This is the subject of much current research and one cannot yet discern the likely outcome. It would take us too far afield to describe these problems here, and the reader who wishes to study them is referred to the current literature (e.g. Feruglio *et al.* 1992, Dreiner and Ross 1993)

Subject to the resolution of this difficulty we can say that the astronomical and cosmological demands on the neutrino decay theory are not incompatible with particle physics, and may even suggest profitable lines of research in that subject.

9
Neutrino Decay and the Ionisation of the Milky Way

9.1 Introduction

In this chapter we introduce the basic idea of our neutrino decay theory. According to this idea (Sciama 1990a) the widespread ionisation of the Milky Way is mainly due to photons emitted by dark matter neutrinos pervading the Galaxy. This idea was proposed because it would immediately solve all the problems described in chapter 5, which arise from the conventional hypothesis that the ionisation sources are bright stars or supernovae. In particular, the ubiquity of the neutrinos could compensate for the small mean free path ($\lesssim 1$ pc) of the ionising photons in the intercloud medium, and their halo distribution could account for the large scale height (~ 670 pc) of the ionised gas in the Reynolds layer.

Of course we can exploit these structural features of the basic idea only if the neutrino decay lifetime τ that would be required is otherwise reasonable. We shall find that we need $\tau \sim 2$ to 3×10^{23} secs. This value is (just) compatible with the lower limits derived in chapter 8, and with certain particle physics theories which are described there. Adopting this lifetime would also have major implications for a large variety of phenomena in astronomy and cosmology other than the ionisation of the Galaxy, and would enable several puzzling problems to be solved.

The most remarkable consequence of the resulting theory is that its domain of validity is highly constrained. As we shall see, it can be correct only if the decay photon energy E_γ, the rest mass m_ν of the decaying neutrinos, and the Hubble constant H_0, each has a value specified with a precision ~ 1 per cent. The required values turn out to be reasonable ones, and in the cases of E_γ and m_ν there exists other observational evidence for these precise values.

We consider first the global aspects of the ionising photon flux in the Galaxy, as derived from Hα data. Then we turn to the local ionisation equilibrium in opaque regions of the intercloud medium. Here a key role is played by the electron density n_c, a quantity which is known observationally mainly from pulsar dispersion measure data. Finally we consider a number of other ionisation phenomena in the Galaxy which can be used to test our basic postulate.

9.2 Implications of the Global Hα Data

The global data on H α emission are particularly valuable because, by virtue of the recombination process which gives rise to the Hα line, the strength of this line directly determines the flux of hydrogen-ionising photons (except in the Galactic plane, where the opacity due to dust leads to Hα photons having a mean free path ~ 2 kpc (Reynolds 1984)). In fact in the circumstances envisaged here on average about 0.46 Hα photons are produced in each recombination following the ionisations due to Lyman-continuum photons (Martin 1988). Thus the Hα data act directly as a photon counter, providing global information throughout the Galaxy and beyond. By contrast, the pulsar dispersion measure, being restricted to line segments of varying length, is a more useful diagnostic for ionisation processes located within a mean free path of the propagating Lyman-continuum photons.

Let us consider the Hα survey carried out by Reynolds (1984) and later summarised by him along with other relevant data (Reynolds 1989b). For convenience we repeat here what we have already said about this survey. The Hα intensity is usually measured in rayleighs R ($1\ R = 10^6$ photons cm^{-2} sec^{-1}). Reynolds found that at Galactic latitudes $|b| > 5$ degrees the Hα intensities I_α generally decrease with increasing $|b|$ in a manner consistent with $I_\alpha \sin|b| = 1R$. He attributed this cosecant law for I_α to the recombining gas having a slablike distribution. Reynolds found that along a vertical line of sight from Earth the total number I of H ionisations on both sides of the Galactic plane $\sim 4 \times 10^6$ cm^{-2} sec^{-1}. Since the slab is opaque to the ionising radiation, it follows that I is also the total effective flux of ionising photons. We are thus able to count these photons by direct observation, despite (or rather because of) the large opacity of the column.

If we neglect for the moment the contribution to I_α from extragalactic photons, and if we assume that its main source is decaying neutrinos in the Galaxy, we can immediately derive an estimate for the decay lifetime τ. This estimate follows from knowledge of the column density of neutrinos needed to account for the extended flat rotation curve of the Galaxy. According to the analysis of Caldwell and Ostriker (1981) the required column density σ at the sun's position is about 5×10^{-2} gm cm^{-2}. We need to convert σ into a column number density N of neutrinos, which requires knowledge of the neutrino rest-mass m_{ν_τ}. Later on we shall solve self-consistently for this rest-mass, but in order to make a start we shall suppose that $m_{\nu_\tau} \sim 30$ eV. As we saw on page 71 this is the value which would follow from assuming that $\Omega_{\nu_\tau} \sim 1$ and $h \sim 0.5$ (the value of h most consistent with constraints coming from estimates of the age of the universe). A factor two uncertainty in m_{ν_τ} would correspond to a factor two uncertainty in N, and so in τ, but we shall see from other considerations that m_{ν_τ} is in fact required to be very close to 30 eV. With this choice for m_{ν_τ} we have that $N \sim 10^{30}$ cm^{-2}.

We now assume that

$$I_\alpha \sim \frac{N}{\tau},$$

thereby neglecting geometrical factors of order unity arising from the transparent parts of the vertical line of sight. With $I_\alpha \sim 4 \times 10^6$ cm^{-2} sec^{-1} and $N \sim 10^{30}$ cm^{-2} we then obtain

$$\tau \sim 2.5 \times 10^{23} \text{ sec},$$

which is the fundamental parameter of the theory.

In view of the various uncertainties in the model we cannot regard our derived value of τ as very precise. Given that m_{ν_τ} will eventually be determined with a precision ~ 1 per cent, an uncertainty of ~ 50 per cent either way in τ would perhaps be reasonable. Our value of τ then has strong implications for the extragalactic flux of ionising photons emitted by the cosmological distribution of neutrinos. This point will be discussed in chapter 11.

9.3 The Local Ionisation Equilibrium

We begin this section by considering an opaque region of the interstellar medium. The density of HI in such a region has been reviewed by Dickey and Lockman (1990). Adopting a number density $n_{\rm HI}$ for the intercloud medium in the range 0.1 to 1 cm^{-3}, we would have a mean free path l for ultra-violet photons with energy $E \geq 13.6$ eV given by $l = (n_{\rm HI}\sigma)^{-1}$, where the ionisation cross-section of hydrogen σ is given by

$$\sigma \sim 7 \times 10^{-18}(13.6/E)^3 \text{ cm}^2,$$

so

$$l \sim (0.5 \text{ to } 0.05)\left(\frac{E}{13.6}\right)^3 \text{ pc}.$$

Thus if one is considering opaque regions much larger than $\sim$ 1 pc, it should be a good approximation to assume that every decay photon produces an ionisation locally, to be followed by a recombination. Equating the ionisation and recombination rates, we would have

$$\frac{n_\nu}{\tau} = \alpha n_e^2,$$

where, as usual, the recombination rate α excludes recombinations directly to the ground state, since such recombinations yield photons which would simply produce further ionisations.

We must first check that this ionisation equilibrium can be achieved in the time available. The recombination time $t_r \sim (\alpha n_e)^{-1}$. Since α depends on the electron temperature T (approximately as $T^{-0.75}$ (Osterbrock 1989, pages 16-17)), we need to know this temperature. In the intercloud medium we have $T \sim 10^4$ K (Spitzer 1978), and for this temperature $\alpha \sim 2.6 \times 10^{-13}$ cm^3 sec^{-1} (Osterbrock 1989, page 19). Taking $n_e \sim 0.033$ cm^{-3} from Nordgren *et al.* (1992), we have

$$t_r \sim 2 \times 10^{14} \text{ sec}.$$

This is much less than the age of the Galaxy ($\sim 3 \times 10^{17}$ sec).

We next consider the ionisation time t_i per hydrogen atom. This time is given by

$$\begin{aligned} t_i &\sim n_{\rm HI}/\alpha n_e^2 \\ &\sim 7 \times 10^{14} \text{ to } 7 \times 10^{15} \text{ sec}. \end{aligned}$$

This time is also much less than the age of the Galaxy. It is thus a good approximation to assume that ionisation equilibrium holds in opaque regions. Thus

$$n_e = \left(\frac{n_\nu}{\alpha\tau}\right)^{1/2}$$

and so

$$n_e \sim \left(\frac{n_\nu}{\alpha_o\tau}\right)^{1/2}\left(\frac{T}{T_0}\right)^{0.37}.$$

We see from this simple formula that in opaque regions n_e has four fundamental properties:

1) n_e does not depend explicitly on the total gas density n or the neutral density n_{HI}.

2) n_e depends weakly on T, which itself is fairly constant in the intercloud region (Spitzer 1978).

3) n_e is thus approximately constant in regions where n_ν is approximately constant.

4) Over wider regions $n_e \propto n_\nu^{1/2}$, so that the electron density distribution can be mapped on to the neutrino density distribution, and vice versa, so long as one confines oneself to opaque regions.

We shall see shortly that these properties lead to a number of observational tests of the theory. First we must check whether the relation $n_e = (n_\nu/\alpha\tau)^{1/2}$ does give the value of the electron density derived from analysis of the pulsar dispersion measure data. We have already estimated τ, so it remains to determine the volume density n_ν of neutrinos. We know the column density N of neutrinos at the sun, so we need to estimate the scale height l of the neutrino distribution. We begin by assuming that l has the Caldwell-Ostriker value of $\sim$ 8 kpc. Allowing for the fact that the column density refers to a full line at right angles to the galactic plane, we find that

$$n_\nu \sim 2 \times 10^7 \text{ cm}^{-3}$$

near the sun. We now assume that $\alpha \sim 2.6 \times 10^{-13}$ cm^3 sec, and obtain

$$n_e \sim 0.017 \text{ cm}^{-3}.$$

This value of n_e is significantly less than 0.033 cm^{-3}, the mean value derived from the pulsar data (Nordgren *et al.* 1992). In order to obtain agreement between our theory and observation with this value of n_e we would need to increase n_ν by a factor ~ 4, to $\sim 8 \times 10^7$ cm^{-3}. We could achieve this if we could reduce the scale height by a factor ~ 4 to ~ 2 kpc. This would require that the dark halo of our Galaxy is flattened by a factor ~ 4. Allowing for the uncertainties we could reduce this to a factor ~ 3. We here make contact with the discussion on page 7 of flattened halos of galaxies and of the problem of the density of the Oort dark matter near the sun. We saw there that a flattening by a factor ~ 3 would be allowed by the observational data and various theoretical considerations, and would fit in with observations of other galaxies. We therefore regard this requirement of the decaying neutrino theory as an interesting and acceptable prediction.

9.4 The Scale Height of the Free Electron Distribution

Having derived the result that n_e is approximately constant in opaque regions where n_ν is approximately constant, we must understand the observation that the pulsar dispersion measure saturates at a height ~ 700 pc above the Galactic plane. Up to this height we would expect n_ν to be approximately constant, so the natural assumption is that the Galaxy is running out of gas at this height. Clearly when the total intercloud gas density n starts to approach the value $(n_\nu/\alpha\tau)^{1/2}$, we would have $n \sim n_e$ and $n_{\rm HI} \ll n_e$. What this means is that near such a place there are too few atoms to absorb all the photons emitted in a recombination time, and ionisation breakthrough takes place.

We may estimate the scale length for vertical variations of $n_{\rm HI}$ near breakthrough as follows. Writing $n_0 = (n_\nu/\alpha\tau)^{1/2}$ we would have

$$n_{\rm HI} = n - n_0$$

and so

$$\frac{dn_{\rm HI}}{dr} \sim \frac{dn}{dr}.$$

Defining the scale heights $\lambda_{\rm HI}$ and λ_n by $n_{\rm HI}/(dn_{\rm HI}/dr)$ and $n/(dn/dr)$ respectively, we obtain

$$\lambda_{\rm HI} \sim \frac{n_{\rm HI}}{n}\lambda_n.$$

Thus where $n \sim 1.1\ n_0$, say, we would have $\lambda_{\rm HI} \sim 0.1\ \lambda_n$. This means that in the final stages before breakthrough the value of $n_{\rm HI}$ is dropping rapidly and a sharp HI/HII interface would be produced of a type familiar from the edges of classical HII regions. The sharpness of the interface would in practice be somewhat smoothed out by local variations in the total gas density n.

Above the HI/HII interface the mean free path of ionising photons in the intercloud gas will be very long. Accordingly the flux of these photons coming from neutrinos lying above the interface will be large. To this large flux must be added the intergalactic flux of ionising photons. We must ensure that the HI layer observed below the interface is adequately shielded from this flux. To estimate it we note that the vertical flux of decay photons emitted in the opaque layer is $(n_\nu l)/\tau$, where l is the thickness of the layer ($\sim$ 670 pc). Thus the internal flux $\sim 10^6$ cm^{-2} sec^{-1}. Since the total ionising flux $\sim 2 \times 10^6$ cm^{-2} sec^{-1}, the external flux also $\sim 10^6$ cm^{-2} sec^{-1}.

It is important to understand, both here and in other contexts, that a flux of this order is easily absorbed by a relatively modest distribution of clouds, especially in our case when, as we shall see, the photon energy is close to the ionisation threshold of hydrogen. To illustrate what is involved it is convenient to consider first a set of clouds each of which has a gas number density $n \sim 1$ cm^{-3} and a typical size $\sim l$. Then the three quantities l, nl and n^2l will each be numerically equal to l (in different units of course). Each of these quantities has a clear meaning. The dimension l influences the number of clouds along a given line of sight. The quantity nl determines the observability of each cloud, while n^2l determines its capacity to absorb the incident ionising flux.

To illustrate further what is involved it is helpful to consider the statistical properties of clouds located near the sun. These properties have been reviewed by Dickey and Garwood (1989). They give the number of clouds ϕ per unit line-of-sight-distance as a function of cloud column density N. Here $\phi(N)dN$ is the number of clouds per kiloparsec having a column density in the range N to $N + dN$ on the average line of sight at $z = 0$ in the Galactic plane. They find that

$$\phi(N) = 2N^{-2},$$

where N is expressed in units of 10^{20} cm^{-2}.

This convenient formula has to be cut off at low N, since otherwise the total column density $\int N\phi(N)dN$ would diverge at the low end. However, it holds down to $N \sim 10^{-2}$ and possibly down to $N \sim 10^{-3}$. At these low values of N (10^{18} cm^{-2} and 10^{17} cm^{-2}) we are dealing with many clouds per kpc with a column density lying between N and $10N$ (200 and 2000).

Neutral interstellar clouds are known to exist at heights beyond 1 kpc (Dickey and Lockman 1991), and there could easily be, say, of order 10 clouds of column density $\sim 10^{17}$ cm^{-2} at these heights along a vertical line of sight. If such clouds have $n \sim 1$ cm^{-3}, we would have $l \sim 10^{17}$ cm, $nl \sim 10^{17}$ cm^{-2} and $n^2 l \sim 10^{17}$ cm^{-5}. If in addition these clouds each have a kinetic and spin temperature ~ 100 K, their optical depth for 21 cm absorption would only be 5×10^{-4}. Their combined column density along a vertical line of sight would be $\sim 10^{18}$ cm^{-2} which would be undetectable at 21 cm in both absorption and emission.

Nevertheless this layer of clouds would completely absorb an incident ionising flux F of $\sim 10^6$ cm^{-2} sec^{-1}. Balancing ionisations and recombinations in the usual way we would have for complete absorption

$$10\alpha n^2 l = F,$$

or

$$n^2 l = 10^5/\alpha.$$

With $T \sim 100$ K we would have $\alpha \sim 10^{-12}$ cm^{-3} sec and so this relation would be satisfied.

Of course on this picture these clouds are being detected through their H α emission which measures $\alpha \int n_e^2 ds \sim 10^6$ cm^{-2} sec^{-1}. At the same time their contribution $\int n_e ds$ to the pulsar dispersion measure is negligible — it is only 1.5 per cent of the asymptotic dispersion measure which sets in at a height ~ 700 pc.

We conclude from this discussion that the external ionising flux can be easily absorbed by clouds which are observable only by their H α emission. In that case the external flux would not influence appreciably the HI/HII transition produced by the internal flux.

We now consider other observational tests of our theory. There is a test involving the constancy of n_e in limited regions of the

Galaxy, but since the observed constancy is derived only after allowing for localised regions of transparent gas, we defer this discussion until we introduce a more realistic model for the distribution of the interstellar medium. Instead we consider here an observational test of property 4) according to which $n_e \propto n_\nu^{1/2}$ in extended opaque regions of the Galaxy.

9.5 Decaying Neutrinos and the Mass Model of the Galaxy

We begin by considering the radial distribution of the electron density n_e in the Galaxy. A model for this distribution was derived by Lyne, Manchester and Taylor (1985) in their extensive analysis of pulsar dispersion measure data. In their first model they assumed that n_e was independent of galactocentric distance r. They found that this model led to a variation of the derived mean height above the galactic plane of groups of pulsars with the distance r of these groups. Arguing that the mean height is mainly determined by the recoil of pulsars at birth, which should be independent of r, they introduced an r-dependent factor into n_e to smooth away this variation. They thereby derived the following r dependence:

$$n_e = 0.025\left(\frac{2}{1 + r/r_o}\right)\ \mathrm{cm}^{-3}.$$

According to our mapping principle we would then have

$$n_\nu \propto \frac{1}{(1 + r/r_o)^2}.$$

This result is remarkable because we would actually expect to find a factor of this general type arising in the radial distribution of dark matter in the Galaxy, that is, one which is independent of r for small r, and $\propto r^{-2}$ for large r (to give a flat rotation curve for large r).

A detailed discussion of the galactic mass model resulting from this dark matter distribution has been given by Salucci and Sciama (1990). They assumed the luminous matter to be distributed like an exponential thin disc

$$I(r) \propto e^{-r/r_D},$$

of length-scale $r_D = 3.5$ kpc (van der Kruit 1987). Then by using the $n_\nu(r)$ relation in the equations giving the condition for centrifugal equilibrium

$$V^2(r) = V_d^2(r) + V_h^2(r),$$

where $V_d(r)$ and $V_h(r)$ are the disc and halo contributions to the circular velocity, and its first moment

$$V(r)\frac{dV(r)}{dr} = V_d\frac{dV_d(r)}{dr} + V_h(r)\frac{dV_h(r)}{dr},$$

they solved for the radial component of the galactic structure and decomposed the circular velocity into its disc and halo components. They thus arrived at a unique mass model for the Galaxy, with no adjustable parameters.

Salucci and Sciama tested this unique model against the rotation curve of the Galaxy which had already been constructed from a variety of observations by Salucci and Frenk (1989). This rotation curve is constant (~ 220 km sec^{-1}) between R_D and $3R_D$ (e.g. Kulkarni, Blitz and Heiles 1982). At outer radii, as indicated by HI data for external galaxies of similar luminosity and Hubble type, it is very likely to decrease out to $6R_D$ and then to stay constant at about 185 km sec^{-1}. The model derived from the decaying neutrino theory was found to agree very well with this empirical rotation curve. Moreover, the model's value for the luminous to total mass ratio at the optical radius (0.78) is in excellent agreement with the value 0.72 ± 0.05 obtained by techniques that do not assume any particular halo density law (Salucci and Frenk 1989, Persic and Salucci 1988, 1990).

9.6 The Diffuse Ionised Gas near the Sun

Another test of the neutrino decay theory involves its prediction that n_e should be approximately constant in opaque regions small enough for n_ν to be approximately constant. We recall that precisely this result has in fact been obtained by Spitzer and Fitzpatrick (1993) for four warm quiescent opaque regions along the line of sight to the star HD93521, and by Reynolds (1990a) and Sciama (1990c) for the lines of sight to the three pulsars with accurately known distances (page 84). It was pointed out there that this constancy would not be expected if the ionising sources were distant O stars whose photons had to find transparent channels to enable them to reach the lines of sight to the pulsars.

It will also be recalled that this analysis required us to make allowance for the presence in the interstellar medium of large bubbles of very low density. The gas in these bubbles is transparent to the ionising photons, and their electron density is abnormally low. Accordingly, we must modify our previous simple model of a uniform opaque intercloud medium by incorporating these bubbles into a more realistic model.

The constant value of n_e which we obtained for the opaque regions on the lines of sight was 0.05 cm^{-3}. This is nearly twice the mean value of n_e derived for many lines of sight by the dispersion measure data (Nordgren *et al.* 1992). The simplest possible model including the bubbles would then consist of a distribution such that on average about half of each line of sight would lie within a bubble.

A further complication arises from the transparency of the bubbles, in that an abnormally large ionising flux would be incident on the sections of the intercloud region adjacent to a bubble. Thus the electron density would increase as one moved in the opaque intercloud region towards an interface before dropping as one entered the bubble. This means that the value of n_e to be used in the local ionisation equilibrium equation should be less than 0.05 cm^{-3}.

Since about half the volume is occupied by opaque regions and half by bubbles, one has twice as many ionising photons available as in a uniform case. The contribution of the bubble photons to the total column density of n_e across an occupied region depends on the density profile of the gas in the interface and so is model dependent. A rough estimate suggests that to derive the electron density in the deep interior of an occupied region one should divide the mean value of n_e for such a region by $\sim 2^{1/2}$. One would thus arrive at a value ~ 0.035 cm^{-3} for the electron density to be used in the local ionisation equilibrium equation. This is essentially the same as the value 0.033 cm^{-3} derived from the uniform model by Nordgren *et al.* (1992) which we used earlier.

9.7 The Ionisation of Nitrogen in the Interstellar Medium

We saw in chapter 5 that in the interstellar medium as a whole "nitrogen is ionised wherever hydrogen is ionised." We also saw in chapter 6 that the same is true in NGC 891. Since nitrogen has a higher ionisation potential than hydrogen the same opacity and scale height problems arise. We therefore propose (Sciama 1992) to solve these problems by supposing that nitrogen as well as hydrogen is mainly ionised by decay photons. This hypothesis would require that

$$E_\gamma > 14.53 \text{ eV}.$$

Although this constraint on E_γ may not seem much stronger than the one following from the ionisation of hydrogen, we will now see that it is actually of great significance for the consistency of the neutrino decay theory.

The reason for this is the existence of opposing constraints giving an upper limit on E_γ. The first of these constraints which was derived (Sciama 1990a, b) comes from the requirement that the intergalactic flux of ionising photons emitted by the cosmological distribution of neutrinos should not exceed the observational upper limit on the intergalactic flux. This constraint, which is discussed in detail in chapter 11, can be written

$$E_\gamma < 15.1 \text{ eV}.$$

While this is not as numerically firm a constraint as the lower limit, it is remarkable how close together the two limits come.

The second constraint, which was obtained more recently (Sciama 1993a), is derived in the next section. It is

$$E_\gamma < 14.68 \text{ eV}.$$

As we shall see, this upper limit is numerically as firm as the lower limit of 14.53 eV, but it does depend on accepting that decay photons are the solution of the C^0/CO ratio problem described in section 5.8.

If we accept these arguments we see that the domain of validity of the neutrino decay theory is non-zero but extremely small, being spread over less than 1 per cent for E_γ (and so also for m_{ν_τ} if $m_{\nu_{e,\mu}}/m_{\nu_\tau} \ll 1$).

From one point of view we could regard the neutrino decay theory as "determining" the values of these quantities. From another point of view we could argue that it would have been quite easy for the theory to have no domain of validity at all. If the ionisation potential of nitrogen had happened to be, say, 16 eV, then the neutrino decay theory would have been immediately disproved. Whether the survival of the theory at this point adds to its plausibility I leave the reader to determine.

9.8 Ultraviolet Radiation inside Dark Interstellar Clouds

So far in this chapter we have considered the need for a flux of hydrogen-ionising photons widely distributed in intercloud regions of the interstellar medium. We saw in section 5.8 that we may also need an excess flux of far ultraviolet photons in the deep interiors of dense molecular clouds in order to solve the C^0/CO ratio problem by increasing the dissociation rate of CO. It is still uncertain whether conventional sources of such photons would be adequate. We therefore here follow Tarafdar (1991) who considered decaying neutrinos as a possible source.

In discussing this possibility we shall update Tarafdar's treatment in a number of respects (Sciama 1993a). In particular we shall assume that the decay photons can ionise nitrogen, so that $E_\gamma > 14.53$ eV, whereas Tarafdar took $E_\gamma = 13.8$ eV. This change in E_γ is important for a number of reasons. First of all the cross-section for the photodissociation of CO is a sensitive function of photon energy because the main process involved is line absorption into predissociated bound states (van Dishoeck and Black 1988). Secondly, the ionisation potential of CO is 14.0 eV, so the decay photons would now be able to ionise CO as well as dissociate it. Thirdly, the lower limit of 14.53 eV for the decay photon energy lies very close to an upper limit which would follow directly from the hypothesis that decay photons can solve the C^0/CO ratio problem, as we shall now see.

This upper limit follows from consideration of the opacity of the molecular clouds for decay photons. In a cloud of number density, say, 10^3 cm^{-3}, most of the hydrogen is known to be in the form of H_2. If the decay photons were energetic enough to lie in the photodissociation continuum of H_2 the resulting opacity would be very large, since the cross-section σ at threshold is 1.1×10^{-17} cm^2 (Lee Po and Weissler 1952). The corresponding mean free path would be $1/\sigma n_{H_2}$, which $\sim 3 \times 10^{-4}$ pc for $n_{H_2} \sim 10^3$ cm^{-3}. By contrast, the opacity due to line absorption into predissociated bound states of H_2 would be relatively unimportant below threshold because these absorbing lines are well-separated (Glass-Maujean *et al.* 1985). Hence if decay photons are to solve the C^0/CO ratio problem we must impose the condition that their energy E_γ is less than the threshold energy of the photodissociation continuum of H_2 which is 14.68 eV (Huber and Herzberg 1979). We thus require that

$$E_\gamma < 14.68 \text{ eV}.$$

The significance of this constraint for the neutrino decay theory will not be lost on the reader and we return to it later.

We now consider other possible sources of opacity. The CO molecule itself is one such source, but since there is evidence that CO is underabundant by a factor ~ 10 (Frerking *et al.* 1982) we shall assume, with Tarafdar, that the main sources are HI and dust grains. This assumption enabled him to estimate the flux of decay photons in a molecular cloud.

The next step is to consider the equilibrium density of C^0. It is formed mainly by the photodissociation of CO and destroyed mainly by reacting with H_3^+. The dissociation cross-section for photons in the energy range 14.53 to 14.68 eV has been measured by Cook *et al.* (1965). The bandwidth of the decay radiation will be about 1Å if the velocity dispersion of the neutrinos ~ 200 to 300 km sec^{-1} and allowance is made for the rotation of the molecular cloud with the Galaxy (assuming that the neutrino halo is rotating more slowly). For this bandwidth we find a cross-section $\sigma_{CO} \sim 10^{-17}$ cm^2.

The ionisation of CO could also lead to its dissociation via the subsequent reaction

$$CO^+ + e^- \rightarrow C + O.$$

However, as Tarafdar has kindly pointed out to me, this reaction will be dominated by

$$CO^+ + H_2 \rightarrow HCO^+ + H$$

$$HCO^+ + e^- \rightarrow CO + H,$$

since $n_{H_2} \gg n_e$ in a molecular cloud .

The density of H_3^+ is determined by a number of reactions which Tarafdar considered in detail. He concluded that

$$n_{H_3^+} \sim 7 \times 10^{-6} \text{ cm}^{-3}.$$

Equating the formation and destruction rates of C^0 we find, following Tarafdar, but using our derived value for σ_{CO}, that

$$\frac{N(C^0)}{N(CO)} = \frac{0.05}{1 + 1.9 \times 10^{-4} n_H}.$$

This is within a factor $\sim$ 2 of a typical observed ratio of $\sim$ 0.1 (Zmuidzinas *et al.* 1986, Frerking *et al.* 1989) if $n_H \sim 10^3$ cm^{-3}. While this derivation involves a number of uncertainties, it is striking that a straightforward calculation comes so close to the observational requirements. We therefore provisionally accept Tarafdar's proposal that decay photons are the solution to the C^0/CO ratio problem.

If we adopt this proposal we must impose the constraint

$$E_\gamma < 14.68 \text{ eV}.$$

This constraint is close to the one imposed by the argument concerning the intergalactic ionising flux which was mentioned in the last section, and which will be discussed further in chapter 11. However it is a stronger constraint in the sense that it is essentially exact, since it is based on an accurately measured threshold. By contrast the intergalactic argument involves some uncertain quantities (the decay lifetime and the observational constraint on the flux). On the other hand we cannot be sure that the C^0/CO ratio problem is solved by decay photons, so that the dissociation threshold of H_2 may be irrelevant. Nevertheless it is striking that the two constraints are numerically so similar, and it is tempting to assume that this is not a coincidence. We therefore now consider the implications of accepting the H_2 constraint.

By combining it with the N constraint we have

$$14.53 < E_\gamma < 14.68 \text{ eV},$$

where both limits are numerically strict ones (if the underlying assumptions are correct). Equally we can write

$$E_\gamma = 14.605 \pm 0.075 \text{ eV}.$$

Thus E_γ would be determined by these arguments with a precision of one part in 200.

The same precision applies to the derived value of m_{ν_τ} if we continue to assume that $m_{\nu_{e,\mu}}/m_{\nu_\tau} \ll 1$. We would then have

$$m_{\nu_\tau} = 29.21 \pm 0.15 \text{ eV},$$

where again both limits are numerically strict if our assumptions are correct.

We may compare this value for m_{ν_τ} with the one which we derived from the phase space constraint for Galactic neutrinos, assuming no phase mixing at the centre of the Galaxy. We found from this argument (page 73) that

$$m_{\nu_\tau} = 27.6 \pm 1 \text{ eV},$$

the high accuracy in this case coming from the insensitive dependence of m_{ν_τ} on the parameters of the galactic neutrino distribution (velocity dispersion and core radius). This value is slightly lower than the one following from the N and H_2 constraints, but the discrepancy is less than two standard deviations.

This good agreement between the values of $2E_\gamma$ and of m_{ν_τ} as derived from the phase space constraint leads to an upper limit on $m_{\nu_{e,\mu}}$ since

$$E_\gamma = \frac{1}{2} m_{\nu_\tau} \left(1 - \frac{m^2_{\nu_{e,\mu}}}{m^2_{\nu_\tau}}\right).$$

We find that

$$m_{\nu_{e,\mu}} \lesssim 5 \text{ eV}.$$

By comparison, the present laboratory limit on m_{ν_e} is

$$m_{\nu_e} < 9 \text{ eV},$$

while the MSW explanation of the solar neutrino problem would imply that

$$m_{\nu_\mu} \sim 10^{-2} \text{ to } 2 \times 10^{-3} \text{ eV}.$$

(Bahcall 1989, Shi *et al.* 1992). These masses for the different neutrino types may be related by the so-called see-saw model (Yanagida 1978, Gell-Mann, Ramond and Slansky 1979). In this model a matrix governs both the masses of the neutrinos and a large cut-off mass M, such as the grand unification mass at which the weak, electromagnetic and strong interactions become equal.

According to one version of this model, when the mass matrix is diagonalised one obtains

$$m_{\nu_e} : m_{\nu_\mu} : m_{\nu_\tau} = m_u^2/M : m_c^2/M : m_t^2/M,$$

where u, c and t refer to the up, charm and top quarks. The values of the quark masses are somewhat model-dependent, but the standard values (Griffiths 1987) are $m_u \sim 4$ MeV and $m_c \sim 1.1$ GeV. The mass of the top quark is more uncertain, but direct experiments imply that $m_t = 130 \pm 25$ GeV (Ellis, Fogli and Lisi 1992). In this case

$$m_{\nu_e} : m_{\nu_\mu} : m_{\nu_\tau} = 1 : 7.5 \times 10^4 : 1.1 \times 10^9.$$

Likely values of M fall into three possible categories (e.g. Rosen and Gelb 1989):

(1) A large value of the order of 10^{15} GeV,

(2) An intermediate value of the order of 10^{11} to 10^{12} GeV,

(3) A low value ($\ll 1$ GeV).

If $m_{\nu_\tau} = 29$ eV, then $M = 1.4 \times 10^{11}$ GeV, corresponding to the intermediate case. We would then have $m_{\nu_\mu} = 1.1 \times 10^{-3}$ eV and $m_{\nu_e} = 1.4 \times 10^{-8}$ eV. It is striking that the value of m_{ν_μ} derived in this way is compatible with the range of values derived from the MSW explanation of the solar neutrino problem . This concordance between neutrino masses determined cosmologically and from the sun has been known in a general way for some time, and has recently been emphasized by Sciama (1990d), Bludman, Kennedy and Langacker (1992) and Ellis, Lopez and Nanopoulos (1992).

It would be important to attempt an experimental test of our prediction that $m_{\nu_\tau} = 29.21 \pm 0.15$ eV. A possible test would follow from the suggestion of Harari (1989) that if $m_{\nu_\tau} = O(10$ eV) one could search experimentally for $\nu_\tau - \nu_\mu$ oscillations if the associated mixing angle were only a factor of a few lower than the present upper bound, say, 0.01. This suggestion is about to be put to the test by experiments at CERN (CHORUS and NOMAD; Winter 1992, di Lella 1992) and Fermilab (P803, Schneps 1991, 1992). In particular, a properly designed experiment could measure a τ neutrino mass of 29 eV if oscillations are detected. (For a different view of these experiments, see Ellis, Lopez and Nanopoulos 1992).

The value of m_{ν_τ} might also be determined by time-of-flight considerations if neutrinos emitted by a nearby supernova were detected by the next generation of water Cerenkov detectors (Minakata and Nunokawa 1990). This possibility has been studied in detail by Krauss *et al.* (1992). They developed a comprehensive Monte Carlo analysis in order to simulate the neutrino signal from a galactic supernova in such a detector. They showed that it should be possible to measure values of m_{ν_τ} down to $\sim$ 25 eV for a medium luminosity burst.

Finally we consider the implications of this theory for the values of the density parameter Ω and the Hubble constant H_0. Following Sciama (1990 b) we note that ρ_{ν_τ} can be derived directly from m_{ν_τ} independently of H_0 because n_{ν_τ} is known to be $\frac{3}{11}n_\gamma$ and so is independent of H_0. Its numerical value is 112 ± 1 cm^{-3} and so

$$\rho_{\nu_\tau} = (6.2 \pm 0.06) \times 10^{-30} \text{ gm cm}^{-3}.$$

Since for this argument we need to know the total density of the universe we must include the density ρ_b in baryons. This was discussed on page 54 where we saw that

$$\rho_b = (0.3 \pm 0.04) \times 10^{-30} \text{ gm cm}^{-3},$$

which is again independent of H_0. This value for ρ_b is 5 times the uncertainty in ρ_{ν_τ}. Combining ρ_{ν_τ} and ρ_b we have

$$\rho = (6.5 \pm 0.07) \times 10^{-30} \text{ gm cm}^{-3}.$$

The final uncertainty in ρ is 1 per cent.

To compare this value with observation it is helpful to express it in terms of Ωh^2. We have

$$\Omega h^2 = 0.31 \pm 0.003.$$

Observationally $\Omega \geq 0.1$ and cannot be much greater than unity (if the cosmological constant is zero, which we shall assume). Our result for Ωh^2 would imply an age t_u for the universe in the range $(12 \text{ to } 8) \times 10^9$ years as Ω varies from 1 to 0.2. The observed age is controversial, with the lowest permitted value being 12×10^9 years (page 59). Thus the theory could be valid only if $t_u \sim 12 \times 10^9$ years and $\Omega \sim 1$. This value of Ω agrees with that recently derived by Kellermann (1993) from the angular diameter-red shift relation for compact radio sources, although this result needs to be confirmed. There is also a theoretical preference for $\Omega = 1$ since this value is the only one which is independent of time. If Ω is indeed exactly 1, so that the universe conforms to the Einstein-de Sitter model, we would have

$$H_0 = 56.3 \pm 0.5 \text{ km sec}^{-1} \text{ Mpc}^{-1}.$$

With this assumption, we would thus have derived H_0 with a precision of ~ 1 per cent and the age of the universe with a precision of better than 10 per cent.

9.9 The Impact of New Data on the Ionisation of the Galaxy

We now consider (Sciama 1993b) the impact on the neutrino decay theory of the new HST data and analysis of Spitzer and Fitzpatrick (1993) which were briefly discussed in section 5.9.

We saw there that these new results are important for the neutrino decay theory in two ways. First of all Spitzer and Fitzpatrick concluded from their analysis that along the line of sight to the Galactic halo star HD 93521 the ionised gas and the neutral gas are mixed up together in each of the nine absorbing regions which were observed, so that these regions must consist of partially ionised gas. As they stressed, the resulting large opacity of the absorbing regions then makes it difficult to account for the observed level of ionisation. This, of course, is just the problem which the neutrino decay theory was designed to solve (together with the scale height problem). The density of neutrinos inside the opaque regions is the same as that outside, so one would always have sources of ionising photons within one mean free path of each point in the opaque regions. All that is then required to solve the problem is that the decay lifetime should have the appropriate value.

The second important result of Spitzer and Fitzpatrick is that the electron density in the four most undisturbed components is remarkably constant. This constancy is a prediction of the neutrino decay theory, as we have seen. Moreover, if one assumes that C, like S, is negligibly depleted in the components, then this common value of the electron density would be the same as that derived for occupied regions along the lines of sight to the three nearby pulsars with accurately known distances, which were discussed in section 5.5. We may recall here the resulting values of n_e. For the four absorbing components one would have 0.05, 0.055, 0.055 and 0.06 cm^{-3} while from the pulsar data one obtains 0.054, 0.056 and 0.056 cm^{-3}.

Fortunately one can test these ideas further by observing more stars with HST. On the neutrino decay theory one would expect to continue finding $n_e \sim 0.05$ to 0.06 cm^{-3} in quiescent warm regions near the sun. Indeed since one is here working at the 10% level of precision, one could contemplate searching for small variations in n_e in different regions resulting, in the neutrino decay theory, from possible small variations in T, in n_ν and in the skin effect for each component.

10
Neutrino Decay and the Ionisation of Spiral Galaxies

10.1 Introduction

We saw in chapter 6 that some nearby spiral galaxies contain diffuse ionised gas (DIG) reminiscent of the Reynolds layer in our Galaxy. This DIG has been studied in particular detail in NGC 891. It was found difficult to account for the DIG observed in that galaxy several kiloparsecs from its plane in terms of known sources of ionisation. The observers concerned therefore concluded that a new galactic source is required, a conclusion which is reminiscent of the situation prevailing for the Reynolds layer in our Galaxy. In this chapter we examine the hypothesis (Sciama and Salucci 1990) that the new source required is decaying dark matter neutrinos with the same properties as we have already invoked in discussing the Reynolds layer in the previous chapter.

This hypothesis has been criticised by Dettmar and Schulz (1992) on the grounds that the decay photons would not heat the gas to the temperature required to account for the emission line ratios [NII]/ Hα and [SII]/ Hα which they observed. This criticism suffers from the defect that in their calculation they assume that the only heat source for the gas is the decay photons themselves. Since in the decaying neutrino theory E_γ is close to 13.6 eV, it is true that there is not much heat input associated with each ionisation. Indeed this point is relevant to our discussion of the temperature of Lyman α clouds in chapter 11. However, in the present case one would expect that other heating processes should be important. Studies of this problem for the interstellar medium have shown, for example, that the liberation of electrons from dust grains by starlight may be a significant heating mechanism. This point has been made by Sciama and Salucci (quoted in Dettmar 1993) and by Reynolds and Cox (1992).

Another popular heating process invokes the photoelectric emission from some large molecules such as PAH's (polycyclic aromatic hydrocarbons) (d'Hendecourt and Leger 1987, Lepp and Dalgarno 1988, Hollenbach 1989, Verstraete *et al.* 1990, Joblin *et al.* 1992). A discussion of this process in the context of the decaying neutrino theory has been given by Sciama (1993b).

It has also been argued by Dahlem, Dettmar and Hummel (1993) that the decaying neutrino hypothesis cannot be applied to NGC 891 because of the substantial asymmetries observed in the H α emission in the north-south regions and also in the east-west regions. They used the result $n_\nu \propto n_e^2$ to infer that the neutrino distributions would also have to be asymmetric, which would lead to unobserved gravitational disturbances. However, the relation $n_\nu \propto n_e^2$ applies only in regions opaque to the decay photons. Where the gas is transparent n_e is essentially equal to the total gas density n, and emission measures will there depend on both n and the effective path-length l. As we shall see below, the asymmetry occurs in the transparent region, and there is no difficulty in supposing that the gas in this region is asymmetrically distributed.

In addition, Dahlem, Dettmar and Hummel state that "the theory can also not explain the observed correlation of the Hα and radio halo." However, as we pointed out on page 97, this correlation may arise from the venting of gas and cosmic ray electrons into the regions concerned as a result of multiple supernova activity in the disk. We therefore believe that it is worthwhile to apply the decaying neutrino hypothesis to NGC 891.

Our basic procedure is to study first the opaque regions of NGC 891 where we may map the electron distribution on to the neutrino distribution using the relation

$$n_e^2(r, z) \propto n_\nu(r, z).$$

This mapping enables us to relate two sets of observational data which would normally be regarded as unrelated, namely the distribution of Hα emission and the rotation curve of the galaxy (Sciama and Salucci 1990). This relation leads to an observational test of the decaying neutrino hypothesis which is in fact satisfied. It also enables us to determine the distance to NGC 891. This in turn leads to another test of the hypothesis which is also satisfied.

10.2 The (n_e, n_ν) map in NGC 891

Our discussion of this map is much simplified by the convenient fact that NGC 891 is very similar to our own Galaxy in a series of structural properties. In fact, in similar fashion to our Galaxy, the light profile of NGC 891 follows the well-known exponential thin disk law (van der Kruit 1984) and the flat part of the rotation curve is $v(R) = 220$ km sec^{-1} (Sancisi and Allen 1979). Moreover, NGC 891 and our Galaxy are similar in their stellar populations. This implies that, independently of the nature of the dark matter, inside the optical radius the two galaxies have the same dark to visible and mass to light ratios and a similar dark matter density law (Persic and Salucci 1988, 1990).

The first step in our analysis consists of noting (Sciama and Salucci 1990) that at small z the radial distribution of n_e^2 given by Rand, Kulkarni and Hester (1990a, b) can be written in the form

$$n_e^2(r) = 4n_0^2\left(\frac{9.5}{D}\right)\left[\frac{1}{(1+r/a)^2}\right]$$

for $r > 3(D/9.5)$ kpc, where

$$a = 4.5(D/9.5),$$

and D is the distance of NGC 891 in megaparsecs (RKH assumed a distance of 9.5 Mpc). Thus n_e^2 has the same functional dependence on r in NGC 891 as in our Galaxy (Lyne, Manchester and Taylor 1985).

This result is crucial for our decaying neutrino hypothesis. So long as we are considering opaque regions of NGC 891, we expect n_e^2 to be independent of the total gas density and to depend only on the local value of the neutrino density n_ν (so long as the gas temperature has the standard value $\sim 10^4$ K), by the relation $n_e^2 \propto n_\nu$. It follows that in opaque regions n_ν has the same functional dependence on r in NGC 891 as in our Galaxy. Now we saw in the last chapter that this r dependence would lead to the observed rotation curve of our Galaxy if the core radius a of the neutrino distribution ~ 8 kpc, and if n_ν at the sun $\sim 8 \times 10^7$ cm^{-3}. Since the star distribution and the rotation curves of the two galaxies are very similar, it follows that our mapping of n_e^2 on to n_ν in NGC 891 would also be consistent with all the data, if $a \sim 8$ kpc and so

$$D \sim 9.5 \times \frac{8}{4.5} \sim 16.9 \text{ Mpc},$$

and if $n_0^2(9.5/D)$ is equal to $n_e^2(R_\odot)$ for our Galaxy. This latter quantity $\sim 10^{-3}$ cm^{-6}. Moreover, from RKH we have that the former quantity is 1.3×10^{-3} cm^{-6}, so that this condition is satisfied. This agreement also implies that the neutrino halo of NGC 891 is as flattened as that of our Galaxy.

A further check on the consistency of our procedure comes from considering the Tremaine-Gunn phase space constraint on the neutrino distribution in NGC 891. Following our discussion of this constraint on page 73, we would expect that if the phase space density of the neutrinos in the central regions of the galaxy is essentially the maximum value at decoupling then

$$m_\nu^4 \propto \frac{1}{v_0 a^2},$$

where v_0 is the velocity dispersion of the neutrinos. This is a factor $\sqrt{3/2}$ times the contribution of the neutrinos to the circular velocity. Galaxies having the same gradient and amplitude of circular velocity (in this case 0 and 220 km sec^{-1} respectively) have also the same value of v_0 (Persic and Salucci 1988, 1990). Thus for this reason also we would expect the core radius a of the neutrinos to be the same in NGC 891 as in our Galaxy, which again leads to a distance of 16.9 Mpc.

10.3 The Vertical Distribution of n_e in NGC 891

We now use the RKH model fit for $n_e^2(z)$ to test the decaying neutrino hypothesis. This fit is based on data which exclude values of z less than $\sim$ 500 pc (for $D \sim$ 17 Mpc). According to the hypothesis n_e is essentially independent of z at fixed r if the gas is opaque. However, this opacity condition would break down at the height z_0 where the total gas density has dropped down to this constant value. A transition HI/HII layer would occur at this height, above which the gas would be highly ionised. In our Galaxy the pulsar dispersion measure and 21 cm data imply that $z_0 \sim$ 500 pc. For NGC 891 the Hα data of RKH and Dettmar (1990) and the 21 cm data of Sancisi and Allen (1979) indicate that here also $z_0 \sim 500$ pc. We shall therefore assume that the vertical flux of ionising photons above 500 pc is determined by $\int_{500}^{\infty} n_e^2 dz$. We then find from the RKH model that at a distance of 8 kpc from the centre of the galaxy (for $D = 17$ kpc) the vertical flux of ionising photons is 1.5×10^6 cm^{-2} sec^{-1}.

According to the decaying neutrino hypothesis this flux is determined by the column density of dark matter above 500 pc. Since the neutrino densities in the plane and the total column densities are the same for NGC 891 and our Galaxy, we would expect that the column densities above 500 pc would also be the same. This implies that the ionising flux in our Galaxy above 500 pc at the sun's position should also be 1.5×10^6 cm^{-2} sec^{-1}. This condition is indeed satisfied as we saw on page 134.

A further test of these ideas arises from the fact that the column density of dark matter in NGC 891 should itself depend on r in a characteristic way, namely approximately as $1/r$, in order to be consistent with the rotation curve. We would then expect, if the decaying neutrino hypothesis is correct, that the quantity $\int_{500}^{\infty} n_e^2(r, z) dz$ should be roughly proportional to the column density of dark matter and so also roughly proportional to $1/r$. This expectation was discussed by Sciama and Salucci (1990) who showed that it is verified. This means that the observed z dependence of the electron density is compatible with the neutrino column density required by the rotation curve. Such a consistency check is important for ensuring that the scale height of both the electrons and of the neutrinos behaves in a smooth way, so that the radial dependence of the neutrino density in the plane can indeed be used to determine the rotation curve.

10.4 The Distance of NGC 891 and the Value of the Hubble Constant

Our distance of 16.9 Mpc for NGC 891 can be used to obtain a local value for the Hubble constant, that is, one which neglects the peculiar velocities of the galaxy and of the Local Group (Sciama and Salucci 1990). Since the observed velocity of NGC 891 is 712 km sec^{-1} (Tully 1989), we obtain a local value of H_0 of 43 km sec^{-1} Mpc^{-1}. This local value is close to the low edge of the usually quoted range of H_0, as is our predicted global value of 56 km sec^{-1} Mpc^{-1} (page 146). We can use this latter value to determine, in the decaying neutrino theory, the peculiar velocity of NGC 891, and to compare this result with independent determinations.

We first use our global value of H_0 to estimate the so-called kinematical distance D_v of NGC 891, that is, the ideal Hubble velocity corresponding to its distance. We obtain in this way $D_v = 16.9 \times 56 = 946$ km sec^{-1}. This value can be compared with the independent determination of D_v by Faber and Burstein (1988) and Burstein (1989). They used the Tully-Fisher relation to derive the distance of NGC 891 with respect to that of the Coma cluster and then from the observed velocity of the cluster (whose peculiar velocity is relatively small) they derived the kinematical distance of NGC 891. After correcting for Malmquist-bias they obtained for D_v the value 905 km sec^{-1}. The good agreement between the decaying neutrino value for D_v and that derived by Faber and Burstein is a significant test of the decaying neutrino hypothesis because our determination depends on the global value of H_0 required by the hypothesis, whereas the Faber-Burstein method is independent of H_0. We may also conclude that the peculiar velocity of NGC 891 relative to the Local Group is about 250 km sec^{-1}.

10.5 Rotation Curves of Spiral Galaxies as Distance Indicators

Our derivation of the distance to NGC 891 can now be extended to other spiral galaxies in the following way (Salucci and Sciama 1991). Consider a spiral galaxy G with a rotation curve like that of our own Galaxy, and whose stellar disk has a surface density $\propto \exp(-r/r_D)$. If we knew its dark matter density profile we could derive its distance from modelling its rotation curve. However, it is not normally possible to derive this profile from observed quantities alone. What we need is a density profile derived directly from observation, in a manner independent of dynamical considerations. This is precisely what we were able to achieve for NGC 891 by using the decaying neutrino hypothesis to map the observed $n_e^2(r)$ on to $n_\nu(r)$. The requirement that the resulting profile $n_\nu(r)$, and specifically its core radius a, should yield the observed rotation curve led to a determination of the distance to NGC 891.

Salucci and Sciama (1991) then made the assumption that a galaxy of type G would always have a dark matter density profile of the same form, namely $\propto (1 + (r/a))^{-2}$, with the same core radius a ($\sim$ 8 kpc). They chose for this exercise the four NGC galaxies 1300, 5055, 7591 and 7723. The value of a for each galaxy was found by minimising the difference (inside $1.5 < r/r_D < 3$) between the observed rotation curve and that resulting from an exponential thin disk and a dark halo with profile $\propto (1+(r/a))^{-2}$. The requirement that $a \sim 8$ kpc then leads to a distance D for the galaxy.

Since we know the red shift of each galaxy we can then derive the Hubble constant from the relation $H_0 = v_r/D$. Of course, the individual values of H_0 will contain the peculiar velocities of the galaxies, but in the average value of H_0 for the sample of galaxies (to which we can add NGC 891) the peculiar velocities will tend to cancel out. We find in this way from the observational data for the five galaxies that

$$\mathrm{H_0} = 54 \pm 7 \text{ km sec}^{-1} \text{ Mpc}^{-1}.$$

The numerical details are given in the original paper. The quoted error is the internal rms error so that it includes both random observational errors and the contamination of v_r due to the peculiar motions of the galaxies, but it does not include systematic errors. We note that our value for H_0 is in good agreement with that derived from completely different considerations in chapter 9, where we found that $H_0 = 56.3 \pm 0.5$ km sec^{-1} Mpc^{-1}.

We can make a further consistency check of our procedure because kinematical distances are known independently for NGC 1300 and NGC 5055 from the Tully-Fisher relation (Faber and Burstein 1990, Burstein 1990). The values we find, 1150 ± 150 km sec^{-1} and 750 ± 90 km sec^{-1}, respectively, are in good agreement with those obtained by Faber and Burstein, namely 1403 ± 120 km sec^{-1} and 705 ± 60 km sec^{-1}.

10.6 Sharp Edges of HI Disks in Galaxies

We saw in chapter 7 that the HI disks of two galaxies (M33 and NGC 3198) are observed to possess sharp edges. In addition Briggs *et al.* (1980) showed more generally that HI column densities less than 3×10^{18} cm^{-2} are not common at the edges of spiral galaxies. Attempts to explain these effects using a conventional type of extragalactic ionising flux were discussed in chapter 7. These attempts run into difficulties if, as seems likely, the medium contains significant fluctuations in density. In that case the edge would not be as sharp as is observed. In addition, a relatively modest cloud distribution lying above the main HI layer could absorb the incident flux without leading to a sharp edge in the column density of HI. The cloud distribution has to be chosen with some care if it is not to violate observation beyond the edge, but is otherwise a reasonable one.

Here we shall accept this cloudy picture of the medium, and consider the effects produced by photons emitted by neutrinos located within the HI layer. The more general case, involving both internally and externally generated decay photons, has been considered by Corbelli and Salpeter (1993). We saw in chapter 9 that the internal decay photons would be expected to lead to a sharp HI/HII interface at a certain height above the galactic plane. A similar interface should be produced in the plane at a certain distance from the galactic centre. To see this we first recall the ionisation equilibrium equation in an opaque region, namely

$$\frac{n_\nu}{\tau} = \alpha n_e^2.$$

Hence

$$n_e = \left(\frac{n_\nu}{\alpha\tau}\right)^{1/2},$$

which we shall call n_0.

Now consider the situation as one moves either upwards from the galactic plane, or radially in the plane away from the galactic centre. In both cases one would expect the total interstellar gas density n to decrease more rapidly than the neutrino density n_ν, which has an extended halo-type distribution. Thus one would expect to reach a height z_0 or a radius r_0 where

$$n \sim n_0.$$

At such places one would have

$$n \sim n_e$$

and

$$n_{\rm HI} \ll n_e.$$

What this means is that near such a place there are too few atoms to absorb all the photons emitted in a recombination time, and ionisation breakthrough takes place.

Maloney (1992) has calculated the critical column density N_{tot} of hydrogen for values of r just less than r_0, where the sudden drop in $N_{\rm HI}$ has occurred. This calculated value can then be used to estimate the critical value N_c of $N_{\rm HI}$. The key result of the calculation is that the total column density N_{tot} of the gas at r_0 has a universal value, except for a nearly linear dependence on the temperature T of the gas and a weak dependence on the eccentricity e of the neutrino halo. Maloney obtained for our neutrino parameters

$$N_{tot} = \frac{7 \times 10^{19}\, T_4^{0.87}}{(1-e^2)^{1/2}}\ \mathrm{cm}^{-2},$$

where T_4 is the temperature in units of 10^4 K.

Maloney reasoned as follows: For a given total surface density $\Sigma_{\rm H}$ of dark matter neutrinos, the midplane neutrino volume density n_ν is given by

$$n_\nu = \frac{\Sigma_{\rm H}}{2 Z_h m_\nu},$$

where Z_h is the vertical scale height of the dark matter distribution. Assume an oblate spheroidal density distribution

$$\begin{aligned} \rho &= \rho(\zeta^2) \\ &= \frac{\rho_c}{(1+\zeta/a)^2}, \end{aligned}$$

where ρ_c and a are the halo core density and radius, respectively, and $\zeta^2 = R^2 + Z^2/(1-e^2)$, where R, Z are cylindrical co-ordinates. Then the scale height (for radii outside the core radius) is $Z_h \sim (1-e^2)^{1/2} R$. With this approximation for the scale height, the neutrino midplane density is

$$n_\nu \sim \frac{\Sigma_{\rm H}}{2(1-e^2)^{1/2} R m_\nu}.$$

The critical gas density n_0 is then

$$n_0 \sim \left(\frac{\Sigma_H}{2(1-e^2)^{1/2} R \alpha m_\nu \tau} \right)^{1/2}.$$

We now calculate the column density corresponding to this critical midplane density. If we neglect the self-gravity of the gas (generally a good assumption for the relevant column densities) and ignore the change of halo density with Z (which is a good assumption for all but highly flattened halos) then the gas vertical density distribution will be a Gaussian,

$$n_H(Z) = n_H(0) e^{-Z^2/2\sigma_h^2},$$

where the scale height σ_h is given by

$$\sigma_h = \frac{(a^2+R^2)^{1/2}}{(4\pi G \rho_c^2 a^2)^{1/2}} \sigma_{zz}.$$

The scale height can also be written

$$\sigma_h = \frac{\sigma_{zz} V_A}{4G\Sigma_H (f(e))^{1/2}},$$

where V_A is the asymptotic velocity of the halo rotation curve and the function of halo eccentricity $f(e) = \sin^{-1} e(1-e^2)^{1/2} e^{-1}$. The halo asymptotic velocity is given by

$$V_A^2 = 4\pi G \rho_c a^2 f(e).$$

The midplane gas density corresponding to a given total gas column density $N(\mathrm{H})$ is

$$n_H(0) = \frac{N(\mathrm{H})}{(2\pi)^{1/2}\sigma_h}.$$

We thus obtain

$$\frac{n_H(0)}{n_0} = \frac{(1-e^2)^{1/4}(\alpha\, m_\nu \tau)^{1/2} N(\mathrm{H})}{\pi^{1/2}\sigma_{zz}} \frac{4G[\Sigma_H R f(e)]^{1/2}}{V_A}.$$

The halo surface density is $\Sigma_H(R) \sim \pi \rho_c a (1-e^2)^{1/2}/R$ (for $R \gg a$). With this and the equation for V_A, the second factor on the right hand side of the previous equation can be written

$$\frac{4G[\Sigma_H R f(e)]^{1/2}}{V_A} = (4G)^{1/2}(1-e^2)^{1/4},$$

and so we have

$$\frac{n_H(0)}{n_0} = \frac{N(\mathrm{H})}{\sigma_{zz}} \left[\frac{4G(1-e^2)\alpha\, m_\nu \tau}{\pi} \right]^{1/2}.$$

If we define the critical column density N_{tot} to be that for which this ratio is unity, that is, the gas is ionised all the way to the midplane, then this column density is just

$$N_{tot} = \pi^{1/2}[4G(1 - e^2)\alpha\, m_\nu \tau]^{-1/2}\sigma_{zz},$$

which is independent of all galaxy parameters except the halo eccentricity and the gas temperature (which enters into both σ_{zz} (as $T^{1/2}$) and α (as $T^{3/4}$)). This is Maloney's result. For our neutrino parameters he obtained

$$N_{tot} = \frac{7 \times 10^{19}\, T_4^{0.87}}{(1 - e^2)^{1/2}} \text{ cm}^{-2}.$$

We now estimate the column density N_c of HI where the edge begins. Of course this beginning is not well-defined; we shall assume that it occurs where $n = 1.5\, n_0$. The height z_0 of the HI layer is given by $n(z_0) = n_0$. Using the relations given above we easily find that

$$\begin{aligned} N_c &\sim \frac{1}{3} N_{tot} \\ &\sim \frac{2.3 \times 10^{19}\, T_4^{0.87}}{(1 - e^2)^{1/2}} \text{ cm}^{-2}. \end{aligned}$$

Thus the model for sharp edges produced by internal neutrino decay predicts that for different galaxies there is no variation in the cut off column density except for that caused by varying halo eccentricity and varying gas temperature. The eccentricity term is unlikely to vary by more than a factor of ~ 2. The temperature is probably determined by the heating effect of the decay photons themselves plus that due to the ionisation of helium by quasar photons, as in the Lyman α cloud problem (page 164). One might then expect a gas temperature close to 10^4 K, with variations no greater than a factor ~ 2. Moreover a lower temperature and a higher eccentricity have an opposite influence on N_c. One would therefore predict a rather similar critical column density for dwarf galaxies like M33 and large spirals like NGC 3198.

By contrast, models in which the ionisation is produced by the extragalactic radiation field predict that N_c should scale as $(V_A/\Sigma_H)^{1/2}$ (Maloney 1993). This quantity varies by a factor ~ 3 from dwarf galaxies to large spirals. The difference between the two models arises because for decaying neutrinos the ionising flux is determined by the density of dark matter, which also determines the equilibrium gas densities. In contrast, if the ionising photons are produced by an external source (the integrated extragalactic background) then the photon flux is independent of the local dark matter distribution. This means that it may be possible to discriminate between the two models by looking for the cut off in galaxies with very different halos, that is, very different masses.

The existing data on sharp edges does concern one dwarf galaxy (M33) and one large spiral (NGC 3198). The cut-offs occur in both cases for $N_c \sim 4 \times 10^{19}$ cm^{-2}. The similarity of the cut-offs, and their closeness to the value predicted by the internal neutrino decay theory, are encouraging. A convincing test would require observations of sharp edges in many more galaxies.

A further question concerns the sharpness of the drop in $N_{\rm HI}$. The extragalactic theory has some difficulty in explaining the observed sharpness (Maloney 1993), especially if the medium is non-uniform. We may estimate the scale length of $N_{\rm HI}$ in the neutrino decay theory by using the argument given on page 133. We found there that

$$\lambda_{\rm HI} \sim \frac{n_{\rm HI}}{n} \lambda_n,$$

where $\lambda_{\rm HI}$ and λ_n are the scale lengths of $\rm N_{HI}$ and n respectively. Thus where $n \sim 1.1 n_0$, say, we would have $\lambda_{\rm HI} \sim 0.1\lambda_n$. Our expression for $\lambda_{\rm HI}$ shows that it is itself a rapidly decreasing function as breakthrough is approached and $n_{\rm HI}$ decreases rapidly. The same is true in this region for the scale height of HI above the plane, so that after the drop begins the column density of HI (which is the measured quantity) should exhibit a rapidly steepening decrease. A more detailed calculation has been made by Maloney (1992) who finds that in this region the column density of HI is proportional to $[\ln(n(r)/n_0)]^{3/2}$. This would give a steeper decrease than is observed.

This problem was pointed out by Corbelli and Salpeter (1993), who also suggested a possible remedy, namely the smearing out effect produced by a fluctuating gas density n. By contrast to the external case, the smearing would now be helpful. If ionisation breakthrough does not occur exactly at one radius, but is spread out over a region of dimensions $\sim$ 1 kpc, one could perhaps account for the observed edges using internal decay photons. This dimension would then be one characteristic length of the fluctuating density field. Clearly a detailed calculation of this effect should be carried out, allowing for the "skin effect" mentioned on page 138. Meanwhile I conclude that decaying neutrinos may be able to explain the observed sharp HI edges of galaxies if the interstellar medium in the vicinity of these edges is sufficiently non-uniform. Future observations of sharp edges in other galaxies would provide a valuable test of this theory.

11
The Intergalactic Flux of Ionising Decay Photons

11.1 Introduction

As soon as one calculates the intergalactic flux F_{ext} of ionising photons coming from the decay of cosmological neutrinos one arrives at a striking result which lies at the heart of the neutrino decay theory. All we have to do is to compare F_{ext} with the ionising flux F_G in our Galaxy which we are assuming to come mainly from decaying neutrinos. We know from our earlier discussions that $F_{ext} \lesssim 6 \times 10^5$ cm^{-2} sec^{-1} at $z = 0$, while $F_G \gtrsim 3.4 \times 10^6$ cm^{-2} sec^{-1}. Thus $F_{ext}/F_G \lesssim 1/6$. On the other hand, the column density σ_G of dark matter in the Galaxy required to account for its rotation curve $\sim 5 \times 10^{-2}$ gm cm^{-2}, while the column density of dark matter in the universe σ for $\Omega_\nu \sim 1$ is about 0.2 h gm cm^{-2}. Thus if $h \sim 0.5$ (as would follow from arguments concerning the age of the universe) we would have $\sigma/\sigma_G \sim 2$. The decaying neutrino theory thus immediately faces a discrepancy of at least a factor ~ 12.

We can resolve this discrepancy by taking into account the redshift of the cosmological decay photons, which eventually reduces their energy below the ionisation potential of hydrogen. If we write

$$E_\gamma = (13.6 + \epsilon) \text{ eV},$$

where E_γ is the energy of a decay photon in the rest-frame of the decaying neutrino, then to obtain F_{ext} we must introduce an additional factor of $\epsilon/13.6$ (if this factor is significantly less than 1). We can thus achieve the required upper limit on F_{ext} by taking

$$\epsilon \lesssim 1.1.$$

This is a striking result for three reasons. First of all it implies that

$$E_\gamma \lesssim 14.7 \text{ eV}.$$

On the other hand we saw on page 139 that we also require the decay photons to be able to ionise nitrogen, which implies that

$$E_\gamma \geq 14.53 \text{ eV}.$$

It is remarkable that a solution is just possible, but only if F_{ext} is actually close to 6×10^5 cm^{-2} sec^{-1}.

Of course, the upper limit of 14.7 eV is not precise, since it depends on measured fluxes and column densities which have a combined uncertainty of, say, 50%. Note, however, that this fractional uncertainty refers to ϵ and not to E_γ. If we suppose that $\epsilon < 1.5$ instead of 1.1, then $E_\gamma < 15.1$ eV. Again a solution would be just possible, but only if E_γ is constrained with a precision of order 2 per cent. The reason for this remarkable result is that to obtain E_γ we must add to the uncertain ϵ a much larger quantity (13.6) which is known with high precision.

Secondly we can derive m_{ν_τ} from E_γ immediately if $m_{\nu_{e,\mu}} \ll m_{\nu_\tau}$. We would then have

$$\begin{aligned} m_{\nu_\tau} &= 2E_\gamma \\ &\sim 30 \text{ eV}. \end{aligned}$$

But this is just the value m_{ν_τ} needs to have if $\Omega_{\nu_\tau} \sim 1$ with h close to 0.5 (page 71), the value which was our starting-point. It is also close to the value of m_{ν_τ} which we derived from the phase space constraint for our Galaxy (page 73). I believe that this self-consistency of the theory is truly amazing. Of course, if the theory is correct there is no miracle: F_{ext} is determined by the actual value of ϵ, and N is ionised simply because it can be for this value of ϵ. But if the theory is false then we are faced with two numerical coincidences which have no actual physical significance whatsoever.

Thirdly our result gives the same constraint as one which was derived in chapter 9. We saw there that if decay photons provide the solution of the C^0/CO ratio problem in the dense molecular clouds of our Galaxy, then we require that

$$E_\gamma < 14.68 \text{ eV}.$$

This would be a firm upper limit, since it corresponds to an accurately measured quantity, namely the photodissociation threshold of H_2. The similarity of this upper limit to the one derived from the intergalactic ionising flux lends support to the neutrino decay theory.

11.2 The Red Shift Dependence of F_{ext}

We now compare F_{ext} (or the implied ionisation rate ζ per H atom) as a function of red shift with the observational estimates of ζ given in table 7.1, recalling that the integrated quasar flux may be unable to account for these estimates at $z \sim 0$, 2 and 4. At $z = 0$ we have

$$F_{ext} = \frac{n_{\nu_\tau}}{\tau} \frac{c}{H_0} \frac{\epsilon}{13.6},$$

where n_{ν_τ} is the present cosmological tau neutrino density. Now we know that $n_{\nu_\tau} \sim 112$ cm^{-3} (page 70) independently of m_{ν_τ} and H_0. From our discussion of the ionisation of the Milky Way we have $\tau \sim 3 \times 10^{23}$ sec, and from our discussion of the age of the universe (page 146) we have $H_0 \sim 56$ km sec^{-1} Mpc^{-1}. We then find that $F_{ext} \sim 6 \times 10^5$ cm^{-2} sec^{-1} for $\epsilon \sim 1$.

For $z > 0$ we must allow for the change in $n_{\nu_\tau}(\propto (1+z)^3)$ and H. We have already seen that the decaying neutrino theory requires the universe to be close to the Einstein-de Sitter model. In this model $H \propto (1+z)^{3/2}$. Hence

$$F_{ext} \propto (1+z)^{3/2}.$$

For large z we must take into account the absorption of ionising photons by Lyman α clouds along the line of sight. According to Madau (1992), at $z \sim 4$ this absorption would reduce F by a factor ~ 2. We must also allow for the fact, explained in section 3 of this chapter, that in the decaying neutrino theory the actual values of F_{ext} and ζ derived from the proximity effect in Lyman α clouds will be a factor ~ 2 to 3 greater than the lower limit derived previously. Finally we note that a more accurate calculation of the Gunn-Peterson effect in the decaying neutrino theory has been made by Dodelson and Jubas (1992, 1993). The comparison between this theory and observation in terms of the H ionisation rate ζ is then given in table 11.1.

We see that, given the uncertainties, the theory is compatible with observation at all three red shifts and that decay photons could be mainly responsible for the ionisation of both the general intergalactic medium and individual Lyman α clouds.

Table 11.1.

z	0	2	4
ζ_{obs} (10^{-12} sec^{-1})	≤ 3.6	≥ 7.4 to 11 ≤ 40	≥ 17
ζ_{theor} (10^{-12} sec^{-1})	3.6	18	20

11.3 The Ionisation and Heating of Lyman α Clouds

We have seen that when one uses the proximity effect to derive the flux of ionising photons incident on a Lyman α cloud at a redshift $\sim$ 2 to 3, one arrives at a value somewhat greater than current estimates of the integrated quasar flux. The existence of this discrepancy is generally recognised, and it is usually resolved by introducing new sources of ionising photons such as young galaxies at red shifts exceeding 5. Since the required flux has a value close to that predicted by the decaying neutrino theory without any adjustable parameters, we shall consider here the implications of assuming that decay photons are mainly responsible for the ionisation of hydrogen in Lyman α clouds.

It may become possible to test this assumption by exploiting the large difference between the spectrum of the redshifted intergalactic decay photons and that of any likely more conventional type of astrophysical source. The ionising decay photons would have energies concentrated in the narrow band between 13.6 and $\sim$ 15 eV, whereas conventional spectra would be expected to extend to higher photon energies. This difference could give rise to a number of potentially observable effects. We consider here three of these effects, namely, those concerning the heating of the clouds, the derivation of the ionising flux from the proximity effect, and the ionisation state of He in the Lyman α clouds.

The heating of the clouds in the presence of decay photons has been considered by Sciama (1991a), Rees (1990) and Maloney (1992). The heating due to the decay photons alone would result in a temperature $\sim 10^4$ K, the exact value depending on E_γ. However, as pointed out by Rees and by Maloney, the ionisation of HeI and possibly of HeII by the integrated quasar flux would produce higher energy photoelectrons, and would lead to greater temperatures. The resulting temperature depends on the quasar flux, the density of the clouds, and whether the ionisation of HeII is important. One arrives in this way at a temperature range ~ 1.5 to 3×10^4 K. By contrast if the entire ionising flux has an extended spectrum one obtains a temperature at the high end of this range (Black 1981, Ikeuchi *et al.* 1989).

The observational situation is unfortunately controversial at the moment. Early measurements of the widths of the Lyman α lines had indicated a temperature close to 3×10^4 K. More recently Pettini *et al.* (1990) claimed to have resolved these lines into narrower components, and concluded that the true temperature of Lyman α clouds was 10^4 K or less. This conclusion was challenged by Carswell *et al.* (1991) who maintained that the higher value of the temperature is the correct one. The controversy involves difficult questions of data analysis and has been discussed by Peacock (1991) and at a mini-workshop held at ESO in February 1991 (Shaver, Wampler and Wolfe 1991). Further contributions have been made by Donahue and Shull (1991) and Rauch *et al.* (1993). At the time of writing the controversy is unresolved.

We next consider the derivation of the ionising flux from the proximity effect. As Rees (1990) has pointed out, if this effect is due to the additional ionisation from the quasar, one must allow for the fact that in the neutrino decay theory the quasar has a harder spectrum than the background. The nearby clouds will then be appreciably heated and so will expand. Both these effects would lead to a reduction of the HI concentration in the clouds over and above the direct ionising action of the quasar flux. As a result the derived value of the background flux would have to be increased, by a factor of order 2 to 3 according to Rees. One would then conclude that, according to the neutrino decay theory, the proximity effect implies that $J_{-21} \gtrsim 2$ to 3.

In terms of the ionisation rate ζ one would have $\zeta \gtrsim 7.4$ to 11×10^{-12} sec^{-1}. By comparison, the direct prediction of the neutrino decay theory is $\zeta \sim 18 \times 10^{-12}$ sec^{-1}. The approximate agreement of this theoretical prediction with the derived lower limit of

11×10^{-12} sec^{-1} requires us to make one important change in our previous discussion (Sciama 1991a). At that time it was thought that the background of decay photons was significantly greater than the value of J derived from the proximity effect. This would have meant that the effect could not be due to the excess ionisation from the quasar, and might perhaps have an environmental origin. This conclusion seemed to be supported at the time by the result of Lu *et al.* (1991) that there was no significant correlation observed between the luminosity of the quasar and the strength of the proximity effect.

This conclusion must now be changed for two reasons. First of all the increase in the value of J implied by the proximity effect, if allowance is made for the extra heating due to the quasar in the neutrino decay theory, brings it into closer accord with the neutrino decay value. Secondly, Bechtold (1993) does find a correlation between the luminosity of the quasar and the strength of the proximity effect. Therefore we may conclude after all that in the neutrino decay theory this effect is probably due to excess ionisation from the quasar.

This conclusion may be supported by the recent observation of Moller and Kjaergaard (1992) that the quasar Q 2224 - 408A ($z = 2.331$) does not reduce the number of nearby Lyman α clouds along a line of sight passing alongside and close to the quasar. If the proximity effect had an environmental origin one would expect to observe such a reduction. However, if the effect is due to the ionising flux of the nearby quasar, one could appeal to existing models in which this flux is highly anisotropic, being associated with a jet from the quasar. The observation could then be explained if only about 0.2 of the forward directed ionising flux is emitted at right angles to the line of sight (although there does exist an alternative explanation in terms of quasar variability).

A third consequence of the restricted spectrum of the decay photons has been investigated by Miralda-Escude and Ostriker (1992). This concerns various effects associated with the ionisation state of He in the clouds. The decay photons are not energetic enough to ionise He, but the integrated quasar flux would be able to do so. Miralda-Escude and Ostriker made a detailed calculation of the resulting ionisation state of He, making full allowance for the absorption of the quasar flux by the He in the clouds. Much of this absorption comes from Lyman limit systems, in which Lyman continuum absorption is detectable (Sargent *et al.* 1989). The addition of a large flux of decay photons to the quasar flux would strongly increase the fraction of neutral He in these systems, because the resulting increase in the electron density coming from the ionisation of H would greatly enhance the recombination rate of ionised He. A variety of absorption effects would be associated with this increased neutral He density, most of which still need to be searched for. However, Miralda-Escude and Ostriker considered that the existing data already contradict the neutrino decay hypothesis.

I believe that this conclusion is premature. The Lyman limit systems mainly responsible for the He absorption of the quasar flux are thought to be high red shift galaxies (Wolfe 1991). These galaxies would be likely to contain stars, supernovae and hot gas whose radiation contains photons capable of ionising He both once and twice. For example, in the local cloud in which the sun is immersed the ionised helium fraction is the same as the ionised hydrogen fraction (Kimble *et al.* 1992). A similar reduction in the neutral and singly ionised helium fraction in the Lyman limit systems would reduce the absorption of the quasar flux. The increased quasar flux would in turn further reduce the neutral and singly ionised helium fraction. This is the sensitive situation familiar from HI/HII interfaces at the boundaries of classical HII regions, and also from the sharp HI edges discussed on page 133. Thus the Miralda-Escude and Ostriker result is very sensitive to the ionisation model adopted. Future observations of a He Lyman α forest and of the HeII Gunn-Peterson effect should help to clarify this problem.

These considerations are relevant to the Hubble Space Telescope observations of the quasar HS $1700 + 6416$ ($z = 2.72$) which were made by Reimers *et al.* (1992). They observed HeI in several absorbing clouds along the line of sight to this quasar but failed to observe either a HeI 584$\mathring{A}$ forest or HeI 504$\mathring{A}$ continuum absorption. They claimed that these negative results strengthen the conclusion of Miralda-Escude and Ostriker that the lack of visibility of either phenomenon in HS $1700 + 6416$ rules out the decaying neutrino theory. However these authors again neglected the contribution to the single and double ionisation of He in Lyman limit systems by stars, supernovae and hot gas in these systems. Such a contribution, whose existence is itself indicated by the large ratios of N(HI) to N(HeI) found for several Lyman limit systems by Reimers *et al.*, could easily reconcile their observations with the decaying neutrino theory.

12
The Reionisation of the Universe

12.1 Introduction

In this chapter we study the implications of the neutrino decay theory for the reionisation of the universe and the consequent suppression of fluctuations in the microwave background. We saw on page 48 that we expect the early high temperature universe to have become neutral at a red shift ~ 1000, when it had cooled down to a temperature ~ 3000 K. On the other hand we know from considerations of the Gunn-Peterson effect that the intergalactic medium is highly ionised at redshifts between zero and 4.9. The questions then arise, at what red shift between 4.9 and 1000 did the reionisation occur, and by what process?

These questions, and the general thermal history of the universe, have been much discussed. They are obviously relevant to our understanding of the processes of galaxy formation. In addition it has long been realised that they play a crucial role in determining the present anisotropy $\Delta T/T$ of the microwave background on small angular scales. As has often been discussed (e.g. Efstathiou 1988 and references cited therein), if the post-recombination universe had been reionised so early that its optical depth for Thomson scattering exceeded unity, then the $\Delta T/T$ induced by fluctuations associated with galaxy formation after recombination at $z \sim 1000$ would have been severely attenuated by $z = 0$. This is an important question because the present stringent observational limits on $\Delta T/T$ at small angular scales would impose severe constraints on several theories of galaxy formation in the absence of a scattering screen. In addition the recent much publicised positive results from COBE (Smoot *et al.* 1992) on fluctuations at large angular scales (> 10 degrees) have lent urgency to theoretical studies of these problems.

A related question concerns the positive contribution to $\Delta T/T$

arising from fluctuations in the reionised gas itself. This question has been discussed by Kaiser (1984), Peebles (1987), Efstathiou and Bond (1987), Efstathiou (1988), Ostriker and Vishniac (1986), Vishniac (1987) and Scott, Rees and Sciama (1991). The linear contribution to $\Delta T/T$ is usually small, but Ostriker and Vishniac made the important point that non-linear effects, arising from the coupling between fluctuations in the electron density n_e and the velocity v of the reionised gas, could give rise to larger values of $\Delta T/T$ on the scale of seconds or minutes of arc than the primordial ones which the gas is suppressing. Vishniac (1987) concluded that "attempts to circumvent small-scale temperature fluctuation limits by appealing to an epoch of reionisation do not work."

The scattering effects resulting from the reionisation due to decaying dark matter particles have been considered by Rephaeli and Szalay (1981), Salati and Wallet (1984), Dorosheva and Nasel'skij (1987), Nasel'skij and Polnarev (1987), Nasel'skij, Novikov and Reznitsky (1987), Asselin *et al.* (1988) and Scott, Rees and Sciama (1991). Here we follow closely the last authors, whose calculations were adapted to the version of the decaying neutrino theory which is described in this book. According to this theory, after recombination has occurred the ionisation level rises gradually with time. Our task will be to consider the extent to which the primary radiation pattern will be erased by the new scattering screen, and to calculate the expected microwave background anisotropies, paying particular attention to the form and amplitude of the power spectrum of matter density fluctuations.

12.2 The Reionisation of the Pregalactic Medium

Our first task is to determine the electron density n_e in the universe as a function of red shift after recombination. To do this we need to know the recombination rate α of hydrogen as a function of red shift. We shall assume here for simplicity that $\alpha \sim T_m^{-1/2}$, so we need to know the matter temperature T_m of the pregalactic gas. This temperature is mainly determined by Compton interactions between free electrons and the cosmic thermal radiation background down to the red shift z_i of ionisation breakthrough, where $n_e \sim n$. We shall therefore assume the simple result that

$$T_m = T_\gamma$$

$$= 2.7\ (1+z)\ \mathrm{K}$$

for $z > z_i$.

The second simplification which we can use when $z > z_i$ is that all the ionising decay photons will be absorbed near their point of creation before being redshifted below the ionisation threshold. We thus have an ionisation equilibrium similar to that in opaque regions of the Milky Way, namely

$$\frac{n_\nu}{\tau} = \alpha n_e^2.$$

We here have $n_\nu(z) = n_\nu(0)(1+z)^3$ and $\alpha(z) = \alpha(0)(1+z)^{-1/2}$. Writing χ_e for the ionisation fraction n_e/n, and expressing n in terms of Ω_b (with $h \sim 0.5$) we obtain

$$\chi_e^2 = 3000\ y^{-1}\left(\frac{\Omega_b}{0.1}\right)^2 (1+z)^{-5/2},$$

where we have parametrised the decay lifetime by

$$\tau = 10^{23} y \text{ secs.}$$

We must now verify the validity of our assumption of ionisation equilibrium by showing that with this value of χ_e one obtains a recombination time less than the expansion timescale for all $z > z_i$. This condition is indeed satisfied for $y \sim 1$, the value which characterises our neutrino decay theory.

Ionisation breakthrough will occur at the red shift z_i where $\chi_e \sim 1$. We find that

$$z_i \sim 24 y^{-2/5}\left(\frac{\Omega_b}{0.1}\right)^{-4/5}.$$

Thus for $\Omega_b \sim 0.1$ and $y \sim 1$ we have

$$z_i \sim 24.$$

Note that z_i exceeds the value ~ 5 at which the Gunn-Peterson effect tells us that the intergalactic medium is completely ionised. Of course this effect tells us more than that, since we are required to account for a very low upper limit on the neutral hydrogen density at $z \sim 5$. We know from our previous discussion of this question that with $y \sim 1$ the flux of decay photons is large enough to satisfy this extra condition. This sufficiency of the photon flux manifests itself here in that z_i is comfortably larger than 5. We note also that, with such early reionisation, there may be a photoionisation instability which could provide the seeds for galaxy formation in the decaying neutrino theory (Hogan 1992, 1993). This seeding might overcome the well-known difficulty that hot dark matter models may produce inadequate power in the galaxy distribution on small length scales.

We now consider the optical depth $\tau(z)$ to Thomson scattering by the pregalactic free electrons. For $z < z_i$ the gas is fully ionised. This case was studied by Hogan, Kaiser and Rees (1982). One combines the relation $n_e(z) = n_e(0)(1+z)^3$ with the cosmological path-length factor which, for the Einstein-de Sitter model we are using here, is given by $dz = dt(1+z)^{5/2}$. One thus obtains

$$\tau(z) \sim 1.6 \times 10^{-3}\left(\frac{\Omega_b}{0.1}\right)\left[(1+z)^{3/2} - 1\right].$$

Hence at breakthrough

$$\tau(z_i) \sim 0.18\, y^{-3/5}\left(\frac{\Omega_b}{0.1}\right)^{-1/5},$$

which is relatively small for $y \sim 1$, $\Omega_b \sim 0.1$.

Similarly, for the contribution $\tau(z)$ to τ from the portion of a line of sight with $z > z_i$, we find that

$$\tau(z) \sim 0.49\, y^{-1/2}\left[(1+z)^{1/4} - (1+z_i)^{1/4}\right],$$

assuming that $T_m = T_r$ in this regime. Since $\tau(z)$ is a slowly varying function of z for $z > z_i$ the scattering screen is thick. Our main concern is with the red shift z_* at which unit optical depth is reached. We find that

$$z_* \sim 330$$

for $y \sim 1$, but this value is sensitive to the recombination rate, which depends on $T_m(z)$. Our simple estimates are based on the assumption that T_m stays close to T_r until breakthrough. This assumption starts to break down, however, as χ_e rises towards unity.

Various authors have made more accurate calculations by numerically integrating the full set of governing equations including the heat input from the decay photons. The most recent computations were made by Dodelson and Jubas (1992) who used the Boltzmann equation for the electron distribution. In Fig. 12.1 we show the results of Scott, Rees and Sciama (1991) for $y = 1.5$, $\Omega_b = 0.1$, $h = 0.5$ and $m_\nu = 27.7$ eV. They obtained for this choice of parameters

$$z_i \sim 27$$

and

$$z_* \sim 400.$$

The matter temperature is calculated to remain close to T_r for all redshifts $z > 80$, but rises to more than 10^3 K at $z \sim z_i$. Also plotted is the integral of the visibility function $e^{-\tau} d\tau/dz$, which gives the cumulative probability of a photon being scattered between a redshift z and the present. This function changes relatively slowly with z, so that the scattering screen is thick.

Because z_* is rather sensitive to Ω_b and to y, it is interesting to consider what increase in y (or decrease in Ω_b) would yield an optical depth less than 1 back to the standard recombination era (that is, $z \sim 10^3$). For $\Omega_b = 0.1$, $z_* \sim 10^3$ would correspond to $y \sim 3.2$ or $\tau \sim 3.2 \times 10^{23}$ secs. For $\Omega_b = 0.03$, the corresponding $y \sim 2.2$ or $\tau \sim 2.2 \times 10^{23}$ secs. For our preferred value $\Omega_b = 0.06$ (page 61) $y \sim 2.7$ or $\tau \sim 2.7 \times 10^{23}$ secs. This value of τ lies just inside the allowed range, namely 2 to 3×10^{23} secs, which we derived earlier from consideration of the ionisation of the Milky Way. Thus the degree to which primordial microwave anisotropies are suppressed by the thick scattering screen is a sensitive function of the lifetime τ. It may soon be possible to limit τ in a useful way from observations of $\Delta T/T$ on angular scales $\lesssim 3$ degrees, as we shall now see.

12.3 The Suppression of Fluctuations in the Microwave Background

We have seen that the reionisation produced by decay photons provides a thick scattering screen for microwave photons propagating to us from the last scattering surface ($z = z_*$). This scattering leads to damping of the microwave fluctuations below a characteristic angular scale. When $z_* < 1000$, the limiting scale of

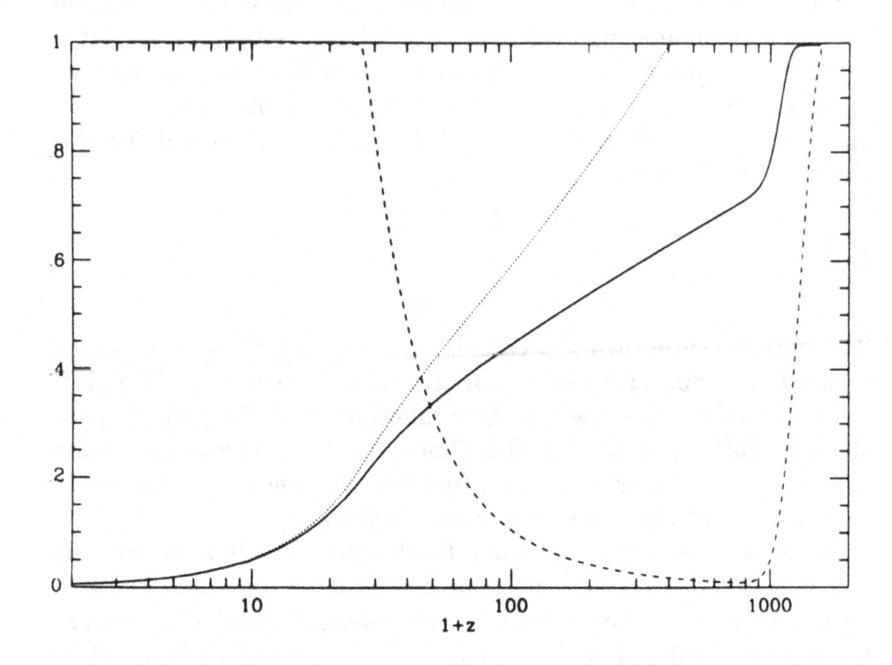

Fig. 12.1 The ionisation history of the IGM for an $\Omega = 1$ universe dominated by decaying neutrinos. The dashed line shows the fractional ionisation, the dotted line shows the optical depth, and the solid line shows the cumulative visibility function (for $\Omega_b = 0.1, h = 0.5, y = 1.5$). It can be seen that $z_i \sim 27, z_* \sim 400$ and that the scattering screen is thick. [From Scott, Rees and Sciama 1991].

the resulting smoothing is given, for causal reasons, by the angle subtended by a horizon-length situated at the redshift z_*. This limiting angle θ_* is of order $(\Omega_0/z_*)^{1/2}$ and since we are adopting $\Omega_0 = 1$ in the neutrino decay theory, we should have

$$\theta_* \sim 3 \text{ degrees.}$$

For standard recombination, on the other hand, θ_* is smaller only by a factor ~ 2. Anisotropies at angular scales larger than this are dominated by fluctuations in the gravitational potential at last scattering, and are independent of the ionisation history (since no causal process can affect scales larger than that of the horizon). It is just such fluctuations which have recently been detected by COBE (Smoot *et al.* 1992). The main difference between the present picture and standard recombination lies at smaller angles. In the standard picture $z_* \sim 1100$, with a thickness (from the visibility function) of $\Delta z/z \sim 1/15$. This corresponds to an angular size of $\theta_* \Delta z/z$ or ~ 8 arc minutes. Below this scale fluctuations would be smeared out because of the superposition of many lumps across the finite last scattering width Δz. This would lead to a break in $\Delta T/T$ on a scale of a few arc minutes. In the decaying neutrino case, however, $\Delta z \sim z_*$ and there is no comparable break scale in $\Delta T/T$ measurements.

We would therefore expect primary anisotropies to be erased to some extent below ~ 3 degrees, but a more precise picture of small angular scale fluctuations would require detailed computations of the induced fluctuations. However, we can obtain a reasonable approximation to the spectrum of erased primary fluctuations without the need for more elaborate calculations. If we look at the visibility curve of Fig. 12.1, which gives the probability for a photon being scattered between now and a redshift z, we notice that about 30% of photons are scattered during the recombination process, and not at smaller redshifts. So we could assume that the primary $\Delta T/T$ pattern is reduced to about 30% of its amplitude at angles below θ_* (i.e. a few degrees) and is unchanged at larger angles.

A better approximation than a sharp step would be to fold in the visibility curve. To do this we can consider each redshift as corresponding to an angular scale given by the horizon size at that epoch (i.e. $\sim z^{-1/2}$). Then the visibility curve of Fig. 12.1 gives one minus the amount by which the primary radiation spectrum is erased at each angular scale. This rather crude method clearly cannot be entirely correct, but probably gives a fair representation of the width and height of the step.

We must also consider microwave background fluctuations which are themselves induced in the reionised gas. It is these fluctuations which Vishniac (1987) claimed were as large as the ones generated at recombination but suppressed by the reionised gas. Scott, Rees and Sciama obtained a much lower value for this Vishniac effect ($\sim 7 \times 10^{-7}$), far below the levels which are currently detectable. One reason for this difference is the different normalisations chosen for the power spectrum of density fluctuations. It would take us too far afield to go into this question here, and we refer the reader to the original paper for a discussion of this and other aspects of the Vishniac effect.

The recent COBE results (Smoot *et al.* 1992) which tell us that $\Delta T/T \sim 10^{-5}$ on angular scales exceeding 10 degrees have invested new urgency into the observational aspects of these problems. Already there have been three reports, more or less simultaneous with COBE, of upper limits on $\Delta T/T$ of 1.8×10^{-5} on an angular scale of 5 degrees (Watson *et al.* 1992), 1.3×10^{-5} on 6 degrees (de Bernardis *et al.* 1992) and 1.4×10^{-5} on 1 degree (Gaier *et al.* 1992). In particular, the results of Gaier *et al.* have led Gorski (1992) to claim that, on the basis of some reasonable assumptions, the observed large-scale streaming motions of galaxies cannot be understood unless the early reionisation of the universe is suppressing $\Delta T/T$ on an angular scale of 1 degree.

The present upper limits at smaller angular scales, based on balloon and terrestrial experiments (e.g. Readhead *et al.* 1989, Fomalont *et al.* 1993), are only a factor 2 or so greater than the value observed by COBE. A relatively small improvement in these experiments should show whether suppression is indeed occurring on angular scales less than 3 degrees. It may be possible quite soon therefore to test the decaying neutrino theory in this way, to measure the red shift of the last scattering surface, and to place strong limits on the decay lifetime.

Part Four
Observational Searches for the Neutrino Decay Line

13
Observational Searches for the Neutrino Decay Line

13.1 Introduction

The evidence which has been accumulated in this book relating to our neutrino decay hypothesis is strong but circumstantial. It is crucially important to test the validity of the hypothesis by attempting to make a direct detection of the postulated radiation. Fortunately the kinematics of the decay imply that the emitted photons are monochromatic, so that the radiation from a given source, if strong enough to be detected, would show up as an unidentified line broadened by the velocity dispersion of the neutrinos in the source. Had the emission possessed a continuous spectrum it would have been much more difficult to distinguish it convincingly from radiation of a conventional origin.

Since the line is predicted to have an energy $E_\gamma \sim 15$ eV, the problem of detectability is tied up with the high opacity of the interstellar medium for radiation of this energy. This problem is a natural one since the opacity is mainly due to the photoionisation of neutral hydrogen, the very process which originally led to the postulate that the decay radiation lies in this energy region. It does mean that care must be taken to choose a suitable observing target.

For example, a number of attempts were made to detect decay photons from dark matter in the Virgo and Coma clusters under the stimulus of the earlier neutrino decay theories of Cowsik (1977) and de Rujula and Glashow (1980). These attempts were made by Shipman and Cowsik (1981), Henry and Feldman (1981) and Holberg and Barber (1985). If opacity problems had not been present, the sensitivity of their detectors, combined with the assumption that the dark matter in these clusters is mainly composed of neutrinos, would have led to lower limits on the decay lifetime τ of order (10^{24} to 10^{25})h sec. Thus these observations, in combination with the dark matter assumption, would have ruled out our choice for τ of 2×10^{23} secs were it not for the opacity problem. This problem means that we must ensure that the photons concerned penetrate the interstellar medium of our Galaxy. However, even the local HI cloud in which the sun is immersed has an optical depth ~ 5 for photons with energies close to 13.6 eV. Thus only a cluster with a large enough red shift could possibly be detectable. The red shift of Virgo is small, and even that of Coma is such that its decay photons would be highly attenuated if $E_\gamma > 13.9$ eV. It is therefore necessary to observe clusters with a larger red shift, despite the difficulties associated with doing this.

A series of such observations, stimulated by the revised neutrino decay theory of Sciama (1990 a), has recently been carried out. We will describe the results of these observations in the next sections of this chapter. We end the chapter by discussing a proposed satellite experiment to search for the decay line from neutrinos within ~ 1 parsec of the sun.

13.2 IUE Observations of the Galaxy Cluster surrounding the Quasar 3C 263

Fabian, Naylor and Sciama (1991) used the IUE database to study the cluster of galaxies surrounding the quasar 3C 263 ($z = 0.646$). This cluster is moderately rich (approximately Abell richness class 2) which means that it is probably not much less massive than the Coma cluster. They searched for extended emission around 3C 263 using a spectrum created by summing a 22.6 arc second region of the line-by-line IUE SWP spectrum produced by the ground station's image processing software (IUS 1PS). The resulting spectrum is shown in Fig. 13.1. No line is observed in the rest-energy range 13.6 to 17 eV, other than geocoronal Lyman α, corresponding to an upper limit of 1.4×10^{-16} ergs cm^{-2} sec^{-1} arc sec^{-2}.

To see what this upper limit would imply for the neutrino decay theory we need to know the column density of dark matter in the IUE aperture. Direct prediction of this column density is not straightforward, since in the absence of detailed x-ray or gravitational lens data the mass profile is poorly known. Approximating

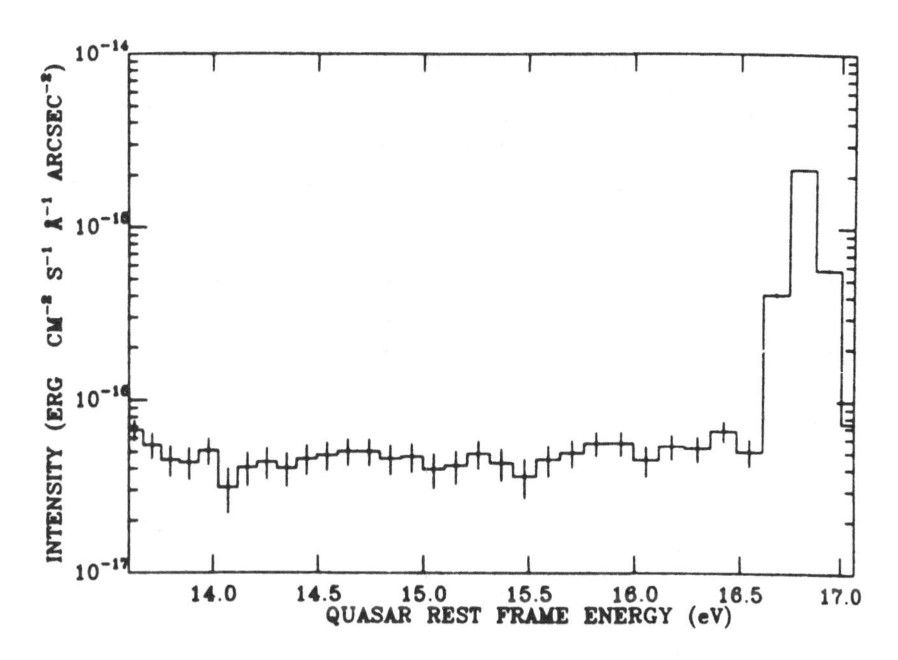

Fig. 13.1 The large-aperture IUE spectrum of the region around 3C 263, in 9.45Å bins. The strong line to the right is geocoronal Lyman α. No point exceeds a line of constant intensity fitted to the data, excluding geocoronal Lyman α, by more than 1.4×10^{-16} ergs cm^{-2} sec^{-1} arc sec^{-2}. [From Fabian, Naylor and Sciama 1991].

the cluster as an isothermal sphere of line of sight velocity dispersion $v = 1000\ v_3$ km sec^{-1} and core radius $r_c = 100\ r_2$ kpc we would expect a column density of $0.7\ v_3/r_2$ gm cm^{-2} (Turner, Ostriker and Gott 1984). When arcs are present the mean surface density Σ within an arc is of order 1 gm cm^{-2}.

At $z = 0.646$, 1 arc sec corresponds to 7.8 kpc so the IUE aperture covers an area of about 80×160 kpc^2 in the centre of the cluster. On the assumption that the cluster surface density consists of neutrinos, the intensity I expected from the predicted decay rate is

$$I = 5 \times 10^{-16} \frac{\Sigma}{1 \text{ gm cm}^{-2}} \text{ ergs cm}^{-2} \text{ sec}^{-1} \text{ arc sec}^{-2}.$$

This corresponds to more than 6 $(\Sigma/1 \text{ gm cm}^{-2})\sigma$ above the noise in the IUE image. One or more of the following then applies: $\Sigma < 0.34$ gm cm^{-2} (at 2σ) in the cluster around 3C 263; the decay lifetime τ exceeds 2×10^{23} secs; or the emission line is absorbed along the line-of-sight.

Absorption is an important possibility, because White *et al.* (1991) have discovered cold, x-ray absorbing clouds in the central regions (inner few 100 kpc) of nearby clusters using x-ray spectra. The surface density of the HI clouds is 10^{21} cm^{-2}, which is very much greater than that required to absorb 15 eV photons ($\sim 10^{17}$ cm^{-2}). The existence of the cold clouds in the intracluster gas demonstrates that cold and hot gas can coexist without evaporation of the clouds taking place (magnetic fields presumably inhibit conduction). If the line is absorbed the energy would be reprocessed into other lines, principally Lyman α emission. Crawford *et al.* (1991) obtained a limit of 7×10^{-13} ergs cm^{-2} sec^{-1} arc sec^{-2} on any diffuse Ly α around 3C 263. This observation would not constrain the neutrino decay theory.

Another major coolant of the gas may be [OII]λ3727Å. Crawford (1991) has placed an upper limit for this line of 1.4×10^{-17} ergs cm^{-2} sec^{-1} arc sec^{-2} at 10 arc sec from the centre. This observation would permit only 10 percent of the line energy to re-emerge as [OII] emission.

We conclude from this discussion that the negative result of this IUE observation is only marginally inconsistent with the neutrino decay theory if the dark matter in the cluster consists mainly of neutrinos. A much stronger statement can be made for the observations of the Abell cluster A665 which were made by the Hopkins Ultraviolet Telescope (HUT) on the Astro I mission.

13.3 The HUT Experiment on Astro I

The Hopkins Ultraviolet Telescope was optimised for spectroscopic observations in the 912 to 1200Å band — the full spectral range covered was 830 to 1850Å in first order. It had a large aperture

(17 × 116 arc sec). In the Astro I flight of December 1990 it made a number of important observations (including that of the nearby white dwarf which we discussed on page 87). Several rich clusters in the range $0.1 \leq z \leq 0.2$ had been selected for observation, but because of spacelab system-level problems, only one of these clusters, Abell 665, was successfully observed.

A665 is an interesting cluster. It is the richest in the Abell catalogue, a strong x-ray emitter, and has the best-measured Sunyaev-Zeldovich effect (Birkinshaw *et al.* 1991). Its red shift is 0.18 and the velocity dispersion based on 33 galaxies is 1200 km sec^{-1} (Oegerle et at 1991). Davidsen *et al.* estimated that the total mass contained in the HUT aperture was 2.9×10^{13} M$_\odot$. If the mass of A665 is predominantly made up of decaying neutrinos, the luminosity of the cluster in the decay line would be

$$L = \frac{4 \times 10^{54}\ \mathrm{M}_{13}}{\tau_{23} E_\gamma / 14\ \mathrm{eV}}\ \mathrm{photons\ sec}^{-1}.$$

The expected width of the line is 11Å *FWHM*. For the mass estimated above, the predicted flux at Earth is

$$\frac{0.089}{\tau_{23} E_\gamma / 14}\ \mathrm{photons\ cm}^{-2}\ \mathrm{sec}^{-1}.$$

In the relevant energy range HUT has an effective area of 8 cm^2 leading to an expected count rate of 0.24 to 0.71 counts sec^{-1} for the range $\tau_{23} \sim 1$ to 3.

A665 was observed with HUT for a total of 1932 secs on 9 December 1990. The spectrum obtained had no features other than those expected from the airglow based on observations of other faint sources and blank fields during the Astro I mission. Fig. 13.2 shows the relevant portion of the data from the full 1932 sec observation, corrected to the rest frame of A665. The positions of the known airglow lines Lyβ, OI at 989Å, Lyγ and Lyδ are indicated. The observed flux of the Lyβ line is 0.58 counts sec^{-1}, essentially equal to the flux expected for the neutrino decay line.

Davidsen *et al.* converted their upper limits on the decay-line intensity for A665 to a lower limit on the lifetime of the neutrino, taking into account the variation in HUT sensitivity with energy and galactic extinction of $E_{B-V} = 0.034$ for a galactic neutral hydrogen column density of 4.7×10^{20} cm^{-2} (Burstein and Heiles 1978), using an extinction law due to Longo *et al.* (1989). Fig. 13.3

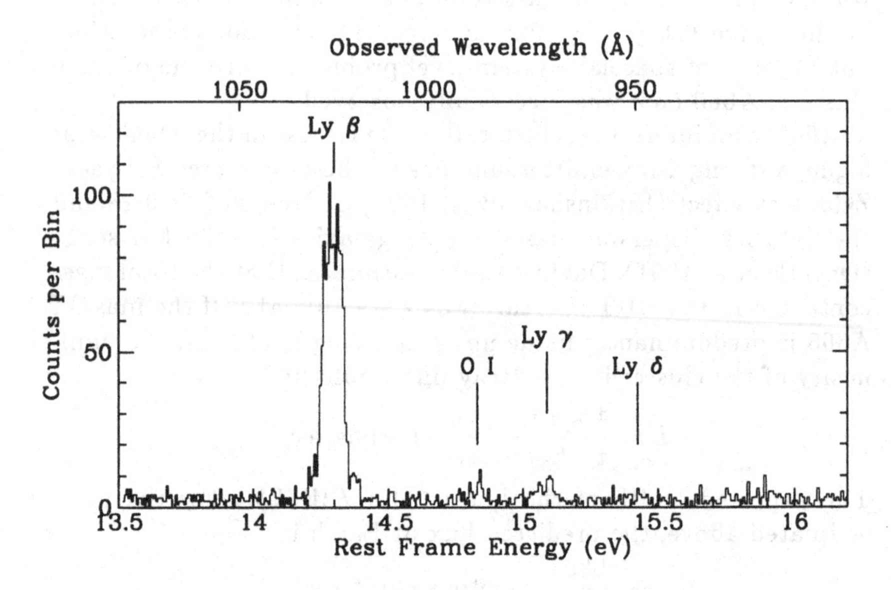

Fig. 13.2 A portion of the A665 HUT spectrum from the full 1932 sec observation covering the rest frame energy 13.6 to 16.1 eV is shown. The ordinate is in observed counts per 0.5 Å bin. Prominent airglow features observed in other HUT spectra are marked. [Reprinted with permission from *Nature* (From Davidsen *et al.* 1991)].

shows the result along with the lifetimes and energies predicted by the decaying neutrino theory.

This figure was compiled before it was appreciated that once-ionised nitrogen is as widespread as ionised hydrogen in both our Galaxy and NGC 891. The allowed range of decay photon energy E_γ is therefore shown as 13.9 to 15 eV. If the decay photons must be able to ionise nitrogen, as we have argued, then the allowed range should be reduced to 14.5 to 15 eV. In this range there is a discrepancy with the neutrino decay theory which varies from a factor 3 to a factor 10, depending on the photon energy. At 14.6 eV, our favoured energy, the discrepancy is about a factor 10.

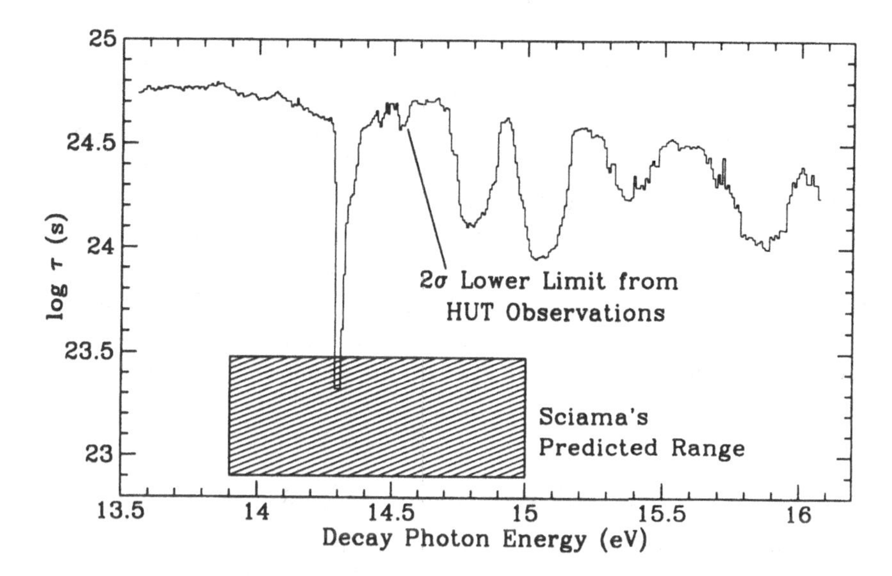

Fig 13.3 The logarithm of the derived 2σ lower limits for the lifetime of decaying neutrinos in seconds is shown as a function of the decay photon energy in eV, assuming that most of the dark matter in A665 consists of neutrinos. The shaded region refers to the predicted range before the ionisation of interstellar nitrogen was attributed to decay photons [Reprinted with permission from *Nature* (From Davidsen *et al.* 1991)].

One explanation for this discrepancy which must be considered is absorption of the decay photons by the cold clouds discovered by White *et al.* (1991) in rich x-ray clusters. This explanation suffers from the problem, already mentioned above in connexion with 3C 263, that one would expect to observe the reprocessed energy of the decay photons, which would be channeled into emission lines produced during the recombination of the gas. Some rich x-ray clusters do show significant optical emission line activity, but according to White *et al.* two clusters which require a large amount of extra x-ray absorption, A478 and A2029, have very weak optical emission lines. It seems unlikely, therefore, that this absorption could be completely responsible for the discrepancy, although it may play a part.

A more likely explanation is that the dark matter in A665 consists mainly of baryons rather than neutrinos. X-ray data for this cluster obtained after the HUT flight by Hughes and Tanaka (1992) indicate that the dark matter in the cluster is more centrally condensed than the visible matter in gas and galaxies (Sciama, Persic and Salucci 1992, 1993). As we discussed on page 62, this would suggest that the dark matter in this cluster is dissipative and so probably baryonic. Of course one would expect some neutrinos to belong to the cluster, but the discrepancy to be explained is only a factor 3 to 10. Thus a significant population of neutrinos could still be present in the cluster. In particular, the galaxies in the cluster could still be donating their dark matter neutrinos.

13.4 Voyager 2 Observations of A85 and A1291

The clusters A85 and A1291, both of which have red shifts of 0.05, were observed by Voyager 2 in February 1991. Over the wavelength range 912 to 1120Å no emission was seen, either line or continuum (Holberg 1992). The resulting upper limits are comparable to those quoted in Holberg and Barber (1985) for the Coma cluster. For A85, the observed upper limit in counts is 0.002 counts sec^{-1}. Thus, one can double the limits (based on 0.001 counts sec^{-1} in Voyager 2) in Table 2 in Holberg and Barber. For example, at 1000Å the limit is 1.8×10^5 photons cm^{-2} sec^{-1}. For A1291 the limit is twice as strong.

The absence of emission at the level expected on the neutrino decay theory could again have two explanations. For a red shift of 0.05 the decay line would be heavily attenuated by the interstellar medium if $E_\gamma > 14.28$ eV. Thus if the photons can ionise nitrogen, so that $E_\gamma > 14.5$ eV, one would not expect to observe the decay line from A85 and A1291. Secondly, if the dark matter in these clusters is mainly baryonic, then again one would not expect to observe the decay lines. In particular, as we have seen, the dark matter in A85 is centrally condensed (Gerbal *et al.* 1992) and so probably baryonic.

13.5 IUE Observations of Clusters of Galaxies

The clusters 0016 + 16 (z = 0.545) and 1558 + 41 (z = 0.610) were observed on July 20 to 21 1991 and low-resolution (SWP) spectra were obtained (Burstein *et al.* 1993) (Fig. 13.4). The uv spectral range covered by the short-wavelength (SWP) camera is approximately 1200 to 2000Å.

The only structures observable in the background were related to well-known sources of noise in the IUE SWP camera or to remnants of previous exposures. No evident features ascribable to extended emission inside the IUE aperture appeared in the wavelength region 1250 to 1500Å. The corresponding upper limit to a possible emission line flux was 10^{-14} erg cm^{-2} sec^{-1}. The expected flux from an isothermal distribution of self-gravitating neutrinos in the clusters with core radius 250 kpc, one dimensional velocity dispersion 1000 km sec^{-1}, and decay lifetime 2×10^{23} sec would be 3 times greater than this upper limit. Again the discrepancy could be due to most of the dark matter in the clusters being baryonic.

13.6 Spectroscopic Search for Highly Red Shifted Decay Lines

An attempt has already been made to find highly red shifted Lyman α emission lines from discrete objects using a spectroscopic study of portions of "blank sky" (Lowenthal *et al.* 1990). Nothing was found, and the resulting upper limits can also be applied to even more highly red shifted neutrino decay lines from discrete objects. The observations were made in the wavelength range 4500 to 7000Å, and since the initial wavelength $\sim$ 850Å, the red shift of the emitting objects would $\gtrsim$5. According to Scott (1993) neutrino masses exceeding $\sim 10^{10}$ $M_\odot$ should have been detected for a decay lifetime $\sim 1.5 \times 10^{23}$ secs.

One must, however, allow for cumulative absorption of the decay photons by Lyman α clouds and Lyman limit systems along the line of sight. Various estimates have been made of the amount of this absorption. According to the recent study of Madau (1992), a photon close to the Lyman limit emitted at a red shift $\sim$ 4 would experience unit optical depth after traversing a red shift interval

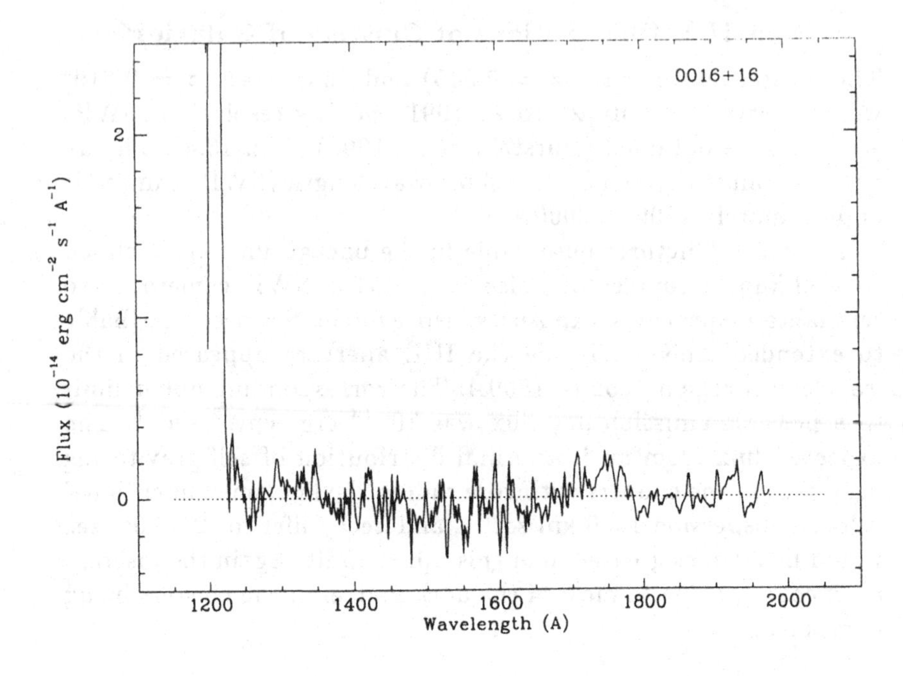

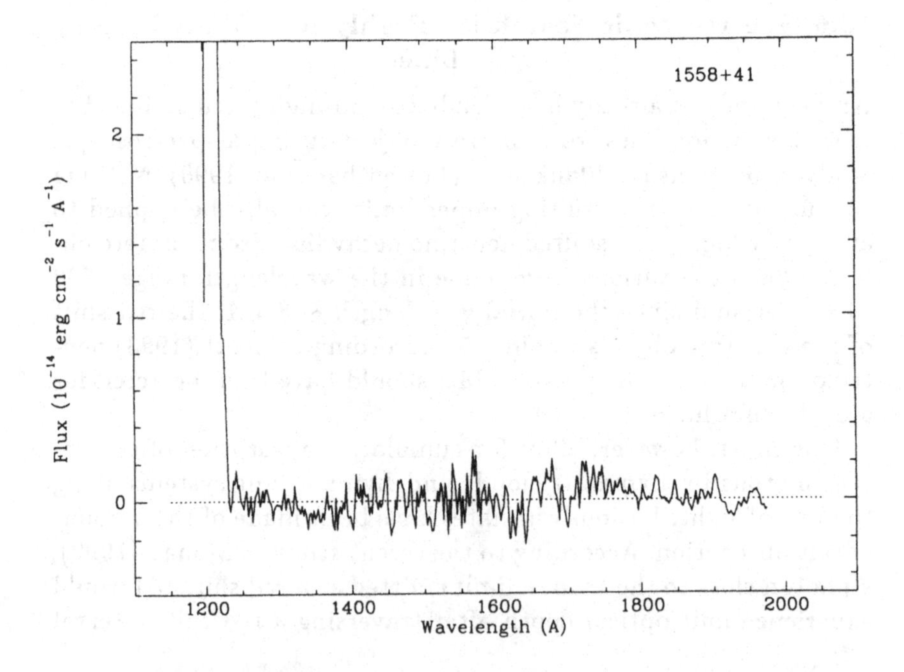

Fig. 13.4 Fully fluxed IUE spectra of the galaxy clusters 0016+16 (z=0.545) and 1558 +41 (z=0.610). The upper limit to an emission line in this wavelength range is 10^{-14} ergs cm^{-2} sec^{-1} . [From Burstein *et al.* 1993].

$\Delta z \sim 0.14$. I estimate that for an emission red shift ~ 5 one would have $\Delta z \sim 0.1$.

One could therefore explain the non-observation of emission lines in regions of "blank sky", if the energy E_γ of a decay photon, which is red shifted as it propagates, remains greater than 13.6 eV for a Δz exceeding 0.1 at $z \sim 5$. In our previous notation this would require that

$$5 \times \frac{\epsilon}{13.6} > 0.1$$

or

$$\epsilon > 0.27.$$

This constraint would certainly be satisfied if, as we are assuming in this book, the decay photons can ionise nitrogen, for that would require

$$\epsilon > 0.9.$$

The negative result of the Lowenthal *et al.* experiment, when applied to the neutrino decay line, would then be expected.

13.7 Proposed Search for a Decay Line from Neutrinos near the Sun

It is clear from the previous discussion that the observation of clusters of galaxies provides a rather uncertain test of the neutrino decay theory. Even if the observed cluster has a large enough red shift to overcome the opacity problem, there is the strong possibility that most of the dark matter in a rich cluster is baryonic. A more decisive test is provided by the neutrinos located close to the sun. It is a crucial feature of the neutrino decay theory that the emissivity near the sun $\sim 2 \times 10^{-16}$ photons cm^{-3} sec^{-1}. Let us estimate the line intensity at the earth resulting from this emission.

This line intensity is controlled by the opacity of the interstellar medium in the vicinity of the sun. We saw on page 86 that the sun lies in a partially neutral cloud whose column density in HI is of order 10^{18} cm^{-2}. This cloud then has an optical depth ~ 5 for photons with energy ~ 15 eV. We are thus able to observe the neutrinos lying within ~ 0.2 of the distance to the edge of the cloud. To determine this distance we need to use the volume density of HI near the sun, which ~ 0.1 cm^{-3}. Unit optical depth for 15 eV photons then corresponds to a physical distance ~ 1 pc. The predicted line strength is then, 600 photons cm^{-2} sec^{-1}.

To this flux must be added a contribution from the neutrinos lying in the hot dilute bubble which surrounds the Local Cloud, which was discussed on page 84. It is difficult to estimate this flux because although the bubble as a whole is transparent to 15 eV photons, it may contain small cold clouds of partially neutral hydrogen which could easily absorb these photons but would be otherwise unobservable. This is a problem which we have already met in other contexts. All we can do here is to obtain an upper limit to the contribution from bubble neutrinos by assuming that the bubble is completely transparent.

The bubble is far from spherical around us, but for a rough approximation we shall assume that its radius ~ 100 pc. The resulting flux on the outer edge of the Local Cloud is then ~ 100 times greater than the flux at the earth from the neutrinos in the Local Cloud itself. We must now allow for the attenuation of the external flux by the Local Cloud. The attenuation is also not spherically symmetric, and we must be content with a rough estimate, especially as the surviving flux depends sensitively on the optical depth. With our adopted values the attenuation is by a factor $\sim 1/100$, so the surviving flux is of the same order as the local flux. The final prediction for the line strength thus lies in the range 600 to 1200 cm^{-2} sec^{-1}, although it could be less than this if the local HI density were, say, 0.15 cm^{-3}.

This predicted flux lies in the sensitivity range of extreme ultraviolet (EUV) detectors. Accordingly an application was made to the European Space Agency (ESA) by Stalio, Bowyer, Sciama and Gimenez to be allotted space on the EURECA 2 platform for an International Diffuse EUV Spectrometer (IDES) to study the diffuse background from 350 to 900Å. This background would be expected to contain detectable ionic emission lines from the Local Cloud interface with the hot bubble and several airglow lines in addition to the postulated neutrino decay line.

This application was accepted by ESA, and IDES has been approved for flight on EURECA 2. Unfortunately ESA's efforts to obtain the necessary funding for flights of EURECA 2 (and 3) have been unsuccessful, and these programmes have recently been put into hibernation. For this reason Spain has decided to add a version of IDES called EURD to their planned Minisatellite Program (MINISAT). The EURD Principal Investigator will be Carmen Morales, and the experiment is scheduled for 1995.

EURD will have essentially the same perfomance as IDES. Since I am more familiar with the latter instrument, I will give a brief description of it here. IDES is a small mass and small size EUV background spectrometer optimised in both sensitivity and spectral resolution over the whole bandpass 350 to 900Å. The instrument will be more sensitive than previous instruments operating in the EUV band, and will also have higher spectral resolution. The IDES optical system consists of two 15 cm diameter Rowland circle spectrographs with holographically ruled 3600 line mm^{-1} diffraction gratings. Each grating views approximately 10×10 degrees of sky through a 100 micron wide slit: one slit is for the 350 to 600Å channel and one for the 550 to 900Å channel. The slit is periodically shuttered for accurate calibration of internal background. Light then strikes the 30 mm diameter imaging microchannel-plate intensified photon counting detectors. Optimum photocathode and grating overcoating will be used.

The IDES mechanical system consists of an optical bench and a separate box for the electronics. The instrument does not need pointing capabilities and is mounted on a fixed plate on the spacecraft viewing the anti-sun direction. Further technical details of the project have been given by Stalio, Bowyer, Sciama and Gimenez (1992). Here we state only that the expected sensitivity after 1000 hours of observation is about 30 photons cm^{-2} sec^{-1}, which is well below our rough prediction for the expected flux of decay photons, and that the expected resolution is about 2 Å. The predicted linewidth also $\sim$ 2Å when allowance is made for the rotation of the Earth around the Galaxy (assuming that the neutrino distribution has a low rotation), as well as for the velocity dispersion of the neutrinos. Thus the line, if detected, may just be resolved. If so, one might be able to determine whether the velocity dispersion in the plane of the Galaxy exceeds that at right angles to the plane, as would be expected if the dark halo is flattened, as discussed on p 7. One might also be able to determine whether the velocity distribution function is triaxial. Thus if the line can be resolved one would be able to make an observational study of the dynamics of the Galactic halo, and measure directly the sun's rotation relative to the neutrino distribution.

If the experiment works as planned it should provide a decisive test of the neutrino decay hypothesis. Absorption is fully allowed for, and it is fundamental for the hypothesis that the emissivity of decay photons near the sun $\sim$ 2 to 3×10^{-16} cm^{-3} sec^{-1}. Thus the neutrino decay theory stands or falls by the result of this experiment. If the theory is correct it should be possible to measure the energy E_γ of the decay line with a precision of about one part in 500. This would provide a sensitive test of our claim that E_γ is already known with a precision of one part in 200 and would lead to an astronomical determination of m_{ν_τ} with a precision of one part in 500, if $m_{\nu_{e,\mu}} \ll m_{\nu_\tau}$.

REFERENCES

Abazov, A. I. et al 1991, *Phys. Rev. Lett.*, **67**, 3332.

Akhmedov, E. K. 1988, *Phys. Lett.*, **B213**, 64.

Akhmedov, E. K., Lanza, A. and Petcov, S. T. 1993, *Phys. Lett.*, **B303**, 85.

Alpher, R. A. and Herman, R. C. 1950, *Rev. Mod. Phys.*, **22**, 153.

Arons, J. and McCray R. 1970, *Astrophys. Lett.*, **5**, 123.

Arons, J. and Wingert, D. W. 1972, *Ap. J.*, **177**, 1.

Ashman, K. M. 1990, *Ap. J.*, **359**, 15.

Ashman, K. M. 1992, *Pub. A.S.P.*, **104**, 1109.

Ashman, K. M. and Carr, B. J. 1988, *M.N.R.A.S.*, **234**, 219.

Ashman, K. M. and Carr, B. J. 1991, *M.N.R.A.S.*, **249**, 13.

Asselin, X., Girardi, G., Salati, P. and Blanchard, A. 1988, *Nucl. Phys.*, **B310**, 669.

Bahcall, J. N., 1989, Neutrino Astrophysics, Cambridge University Press.

Bahcall, J. N., Flynn, C. and Gould, A. 1992, *Ap. J.*, **389**, 234. (BFG)

Bahcall, J. N., Jannuzi, B. T., Schneider, D. P., Hartig, G. F., Bohlin, R. and Junkkarinen V. 1991, *Ap. J.*, **377**, L5.

Bailes, M., Manchester, R. N., Kesteven M. J., Norris, R. P. and Reynolds, J. E. 1990, *Nature*, **343**, 240.

Bailey, M. E. 1982, *M.N.R.A.S.*, **201**, 271.

Bajtlik, S., Duncan, R. C. and Ostriker, J. P. 1988, *Ap. J.*, **327**, 570.

Barcons, X., Fabian, A. C. and Rees, M. J. 1991, *Nature*, **350**, 685.

Barr, S. M., Freire, E. M. and Zee, A. 1990, *Phys. Rev. Lett.*, **65**, 2626.

Bechtold, J. 1993, *Ap. J.* in press.

Bechtold, J., Weymann, R. J., Lin, Z. and Malkan, M. A. 1987, *Ap. J.*, **315**, 180.

Beg, M. A. B., Marciano, W. J. and Ruderman, M. 1978, *Phys. Rev.*, **D17**, 1395.

Begeman, K. 1987, HI Rotation Curves of Spiral Galaxies, Ph. D. Thesis, University of Groningen.

Bergmann, A. G., Petrosian, V. and Lynds, R. 1990, *Ap. J.*, **350**, 23.

Bernstein, J. 1988, Kinetic Theory in the Early Universe, Cambridge University Press.
Bernstein, J., Brown, L. S. and Feinberg, G. 1988, *Rev. Mod. Phys.*, **61**, 25.
Binette, L., Wang, J. C. L., Zuo, L. and Magris, C. G. 1993, *Astron. J.*, **105**, 797.
Binney, J. J. 1992, *Ann. Rev. Astr. Ap.*, **30**, 51.
Binney, J. J., May, A. and Ostriker J. P. 1987, *M.N.R.A.S.*, **226**, 149.
Birkinshaw, M. 1990 in The Cosmic Microwave Background: 25 Years Later, ed. Mandolesi, N. and Vittorio, N., Kluwer, Dordrecht, p. 77.
Birkinshaw, M., Hughes, J. P. and Arnaud, K. A. 1991, *Ap. J.*, **379**, 466.
Black, J. H. 1981, *M.N.R.A.S.*, **197**, 553.
Black, J. H. 1987, in Interstellar Processes, ed. Hollenbach D. J. and Thronson Jr., H. A., Reidel, Dordrecht, p. 731.
Black, J. H. and Dalgarno, A. 1973, *Ap. J.*, **184**, L101.
Blandford, R. D. 1990, *Quart. J.R.A.S.*, **31**, 305.
Blandford, R. D. and Narayan, R. 1992, *Ann. Rev. Astr. Ap.*, **30**, 311.
Bloemen, H. 1991, ed., IAU Symposium No.144, The Interstellar Disk-Halo Connection in Galaxies, Kluwer, Dordrecht.
Bludman, S. A., Kennedy, D. C. and Langacker, P. G. 1992, *Phys. Rev.*, **D45**, 1810.
Bochkarev, N. G. and Sunyaev, R. A. 1977, *Sov. Astr.*, **21**, 542.
Boehm, F. and Vogel, P. 1992, Physics of Massive Neutrinos, Cambridge University Press.
Börner, G. 1990, The Early Universe, Springer, Berlin.
Bond, J. R., Efstathiou, G. and Silk, J. 1980, *Phys. Rev. Lett.*, **45**, 1980.
Bond, J. R. and Szalay, A. S. 1983, *Ap. J.*, **274**, 443.
Bondi, H. and Gold, T. 1948, *M.N.R.A.S.*, **108**, 252.
Bonometto, S. and Pantano, O. 1993, Phys. Rep. in press.
Bowyer, S. 1991, *Ann. Rev. Astr. Ap.*, **29**, 59.
Branch, D. 1992, *Ap. J.*, **392**, 35.
Branch, D. and Tammann, G. A. 1992, *Ann. Rev. Astr. Ap.*, **30**, 359.
Branch, D. and Miller, D. L. 1993, *Ap. J.*, **405**, L5.

Bregman, J. N. and Harrington, J. P. 1986, *Ap. J.*, **309**, 833.
Bridle, A. H. and Venugopal, V. R. 1969, *Nature*, **224**, 545.
Briel, U. G., Henry, J. P. and Bohringer, H. 1992, *Astr. Ap.*, **259**, L31.
Briggs, F. H., Wolfe, A. M., Krumm, N. and Salpeter, E. E. 1980, *Ap. J.*, **238**, 510.
Burstein, D. 1989 private communication.
Burstein, D. and Heiles, C. 1978, *Ap. J.*, **225**, 40.
Burstein, D., Buson, L., Rephaeli, Y., Sarazin, C. and Sciama, D. W. 1993, to be published.
Byrne, J. P. et al 1990, *Phys. Rev. Lett.*, **65**, 289.
Caldwell, J. A. R. and Ostriker, J. P. 1981, *Ap. J.*, **251**, 61.
Campbell, B. A., Davidson, S., Ellis, J. and Olive, K. A. 1991, *Phys. Lett.*, **B256**, 457.
Campbell, B. A., Davidson, S., Ellis, J. and Olive, K. A. 1992, *Astroparticle Phys.*, **1**, 77.
Carlberg, R. G. 1986, *Ap. J.*, **310**, 593.
Carney, B. W., Storm, J. and Jones, R. V. 1992, *Ap. J.*, **386**, 663.
Carr, B. J. 1990, *Comm. Astrophys.*, **14**, 257.
Carr, B. J., Bond, J. R. and Arnett, W. D. 1984, *Ap. J.*, **277**, 445.
Carswell, R. F., Lanzetta, K. M., Parnell, H. C. and Webb, J. K. 1991, *Ap. J.*, **371**, 36.
Carswell, R. F., Whelan, J. A. J., Smith, M. G., Boksenberg, A. and Tytler, D. 1982, *M.N.R.A.S.*, **198**, 91.
Castellani, V. and Degl'Innocenti, G. 1993, *Ap. J.*, **402**, 574.
Chaboyer, B., Deliyannis, C. P., Demarque, P., Pinsonneault, M. H. and Sarajedini, A. 1992a, *Ap. J.*, **388**, 372.
Chaboyer, B., Sarajedini, A., and Demarque, P. 1992b, *Ap. J.*, **394**, 515.
Charlot, S. and Fall, S. M. 1991, *Ap. J.*, **378**, 471.
Cheng, K. P. and Bruhweiler, F. C. 1990, *Ap. J.*, **364**, 573.
Chupp, E. L., Vestrand, W. T. and Reppin, C. 1989, *Phys. Rev. Lett.*, **62**, 505.
Collin-Souffrin, S. 1991, *Astr. Ap.*, **16**, 123.
Cook, G. R., Metzger, P. H. and Ogawa, M. 1965, *Can. J. Phys.*, **43**, 1706.
Corbelli, E., Schneider, S. E. and Salpeter, E. E. 1989, *Astron. J.*, **97**, 390.
Corbelli, E. and Salpeter, E. E. 1993, Ap. J, to be published.

Cordes, J. M., Weisberg, J. M. and Boriakoff, V. 1985, *Ap. J.*, **288**, 221.
Cowie, L. L. 1987, in Interstellar Processes, ed. Hollenbach D. J. and Thronson Jr., H. A., Reidel, Dordrecht, p. 245.
Cowie, L. L., Henriksen, M. and Mushotzky, R. 1987, *Ap. J.*, **317**, 593.
Cowsik, R. 1977, *Phys. Rev. Lett.*, **39**, 784.
Cowsik, R. and McClelland, J. 1972, *Phys. Rev. Lett.*, **29**, 669.
Cowsik, R. and McClelland, J. 1973, *Ap. J.*, **180**, 7.
Cox, D. P. and Reynolds, R. J. 1987, *Ann. Rev. Astr. Ap.*, **25**, 303.
Crawford, C. S. 1991 private communication.
Crawford, C. S., Fabian, A. C., George, I. M. and Naylor, T. 1991, *M.N.R.A.S.*, **248**, 139.
Dahlem, M., Dettmar, R. J. and Hummel, E. 1993, to be published.
Dalgarno, A. and McCray, R. A. 1972, *Ann. Rev. Astr. Ap.*, **10**, 375.
Daly, R. A. and McLaughlin, G. C. 1992, *Ap. J.*, **390**, 423.
Danly, L., Lockman, F. J., Meade, M. R. and Savage, B. D. 1992, *Ap. J. Suppl.*, **81**, 125.
Davidsen, A. F. et al 1991, *Nature*, **351**, 128.
Davidson, K., Kinman, T. D. and Friedman, S. D. 1989, *Astron. J.*, **97**, 1591.
Dearborn, D. S. P., Schramm, D. N. and Hobbs, L. M. 1992, *Ap. J.*, **394**, L61.
de Bernardis, P., Masi, S., Melchiorri, F. and Vittorio, N. 1992, *Ap. J.*, **396**, L57.
Deharveng, J. M., Buat, V. and Bowyer, S. 1990, *Astr. Ap.*, **236**, 351.
Dekel, A. 1991, in Observational Tests of Cosmological Inflation, ed. Shanks, T. et al, Kluwer, Dordrecht, p. 365.
Demarque, P. Deliyannis, C. P. and Sarajedini, A. 1991, in Observational Tests of Cosmological Inflation, ed. Shanks, T. et al, Kluwer, Dordrecht, p. 111.
de Rujula, A. and Glashow, S. L. 1980, *Phys. Rev. Lett.*, **45**, 942.
Desert, F. X., Boulanger, F. and Puget, J. L. 1990, *Astr. Ap.*, **237**, 215.
Dettmar, R. J. 1990, *Astr. Ap.*, **232**, L15.

Dettmar, R. J. 1993, Fund. Cosm. Phys. in press.
Dettmar, R. J. and Schulz, H. 1992, *Astr. Ap.*, **254**, L25.
Dicke, R. H., Peebles, P. J. E., Roll, P. G. and Wilkinson, D. T. 1965, *Ap. J.*, **142**, 414.
Dickey, J. M. and Garwood, R. W. 1989, *Ap. J.*, **341**, 201.
Dickey, J. M. and Lockman, F. J. 1990, *Ann. Rev. Astr. Ap.*, **28**, 215.
di Lella, L. 1992, in Neutrino 92, Granada, Spain.
Dodelson, S. and Jubas, J. M. 1992, *Phys. Rev.*, **D45**, 1076.
Dodelson, S. and Jubas, J. M. 1993, M. N. R. A. S. in press.
Dodelson, S. and Turner, M. S. 1992, *Phys. Rev.*, **D46**, 3372.
Dolgov, A. D. and Zeldovich, Y. B. 1981, *Rev. Mod. Phys.*, **53**, 1.
Donahue, M. and Shull, J. M. 1987, *Ap. J.*, **323**, L13.
Donahue, M. and Shull, J. M. 1991, *Ap. J.*, **383**, 511.
Dorosheva, E. I. and Nasel'skij, P. D. 1987, *Sov. Astr.*, **31**, 1.
Draine, B. T. and Lee, H. M. 1984, *Ap. J.*, **285**, 89.
Dreiner, H. and Ross, G. G. 1993, to be published.
Dubinski, J. and Carlberg, R. G. 1991, *Ap. J.*, **378**, 496.
Dubinski, J. 1992, *Ap. J.*, **401**, 441.
Duncan, R. C., Vishniac, E. T. and Ostriker, J. P. 1991, *Ap. J.*, **368**, L1.
Dydak, F. 1991, in Proc. 25th Int. Conf. on High Energy Physics ed. Rhua, K. K., World Scientific, Singapore.
Efstathiou, G. 1988, in Large Scale Motions in the Universe, ed. Rubin, V. C. and Coyne, G. V., Princeton University Press, p. 299.
Efstathiou, G. and Bond, J. R. 1987, *M.N.R.A.S.*, **227**, 33P.
Ellis, J. 1986, *Phil. Trans. Roy. Soc.*, **320**, 475.
Ellis, J., Fogli, G. L. and Lisi, E. 1992, *Phys. Lett.*, **B274**, 456.
Ellis, J., Lopez, J. L. and Nanopoulos, D. V. 1992, *Phys. Lett.*, **B292**, 189.
Enqvist, K., Masiero, A. and Riotto, A. 1992, *Nucl. Phys.*, **B373**, 95.
Enqvist, K., Olesen, P. and Semikov, V. 1992, *Phys. Rev. Lett.*, **69**, 2157.
Eyles C. J. et al 1991, *Ap. J.*, **376**, 23.
Faber, S. M. and Burstein, D. 1988, in Large Scale Motions in the Universe, ed. Rubin, V. C. and Coyne, G. V., Princeton University Press, p. 116.

Faber, S. M. and Gallagher, J. S. 1979, *Ann. Rev. Astr. Ap.*, **17**, 135.

Fabian, A. C., George, I. M., Miyoshi, S. and Rees, M. J. 1990, *M.N.R.A.S.*, **242**, 14P.

Fabian, A. C., Naylor, T. and Sciama, D. W. 1991, *M.N.R.A.S.*, **249**, 21P.

Fabian, A. C., Nulsen, P. E. J. and Canizares, C. R. 1991, *Astr. Ap. Rev.*, **2**, 191.

Falgarone, E. and Lequeux, J. 1973, *Astr. Ap.*, **25**, 253.

Fall, S. M. and Pei, Y. C. 1993, Ap. J. in press.

von Feilitzsch, F. and Oberauer, L. 1988, *Phys. Lett.*, **B200**, 580.

Ferlet, R. 1981, *Astr. Ap.*, **98**, L1.

Feruglio, F., Giudice, G. F., Masiero, A., Pietroni, M. and Riotto, A. 1993, to be published.

Field, G. B., Goldsmith, D. W. and Habing, H. J. 1969, *Ap. J.*, **155**, L149.

Fischler, W., Giudice, G. F., Leigh, R. G. and Paban, S. 1991, *Phys. Lett.*, **B258**, 45.

Fitchett, M. J. 1990, in Clusters of Galaxies , ed. Oegerle, W. R., Fitchett, M. J. and Danly, L., Cambridge University Press, p. 111.

Flam, F. 1992, *Science*, **258**, 393.

Fomalont, E. B., Partridge, R. B., Lowenthal, J. D. and Windhorst, R. A. 1993, *Ap. J.*, **404**, 8.

Fowler, W. A. 1987, *Quart. J.R.A.S.*, **28**, 87.

Freeman, K. C. 1970, *Ap. J.*, **161**, 802.

Frenk, C. S. 1991, in Observational Tests of Cosmological Inflation, ed. Shanks, T. et al, Kluwer, Dordrecht p. 355.

Frenk, C. S., White, S. D. M., Davis, M. and Efstathiou, G. 1988, *Ap. J.*, **327**, 507.

Frerking, M. A., Langer, W. D. and Wilson, R. W. 1982, *Ap. J.*, **262**, 590.

Frerking, M. A., Keene, J., Blake, G. A. and Phillips, T. G. 1989, *Ap. J.*, **344**, 311.

Frisch, P. C. and York, D. G. 1989, in Extreme Ultra Violet Astronomy, eds. Malina, R. F. and Bowyer, S., Pergamon Press, Oxford.

Gabbiani, F., Masiero, A. and Sciama, D. W. 1991, *Phys. Lett.*, **B259**, 323.

Gaier, T. et al 1992, *Ap. J.*, **398**, L1.
Gell-Mann, M., Ramond, P. and Slansky, R. 1979, in Proc. Supergravity Workshop eds. van Nieuwenhuizen, P. and Freedman, D. Z., North Holland, Amsterdam, p. 315.
Gerbal, D., Durret, F., Lima-Neto, G. and Lachieze-Rey, M. 1992, *Astr. Ap.*, **253**, 77.
Gerhard, O. E. and Spergel, D. N. 1992, *Ap. J.*, **389**, L9.
Gerstein, S. S. and Zeldovich, Y. B. 1966, *Zh. Exsp. Teor. Fiz. Pis'ma Lett.*, **4**, 174.
Giallongo, E., Cristiani, S. and Trevese, D. 1992, *Ap. J.*, **398**, L9.
Gilmore, G. 1991 private communication.
Gilmore, G. 1992, in Evolution of Interstellar Matter and Dynamics of Galaxies, ed. Palous, J., Burton, W. B. and Lindblad, P. O., Cambridge University Press, p.318.
Glashow, S. L., Iliopoulos, J. and Maiani, L. 1970, *Phys. Rev.*, **D2**, 1285.
Glass-Maujean, M., Breton, J. and Guyon, P. M. 1985, *J. Chem. Phys.*, **83**, 1468.
Glassgold, A. E. and Langer, W. D. 1974, *Ap. J.*, **193**, 73.
Gorski, K. M. 1992, *Ap. J.*, **398**, L5.
Gredel, R., Lepp, J. and Dalgarno, A. 1987, *Ap. J.*, **323**, L137.
Green, J., Jelinsky, P. and Bowyer, S. 1990, *Ap. J.*, **359**, 499.
Griest, K. 1991, *Ap. J.*, **366**, 412.
Griest, K. and Roszkowski, L. 1992, *Ap. J.*, **46**, 3372.
Griffiths, D. 1987, Introduction to Elementary Particles, Wiley, New York.
Grossman, S. A. and Narayan, R. 1989, *Ap. J.*, **344**, 637.
Gry, C., York, D. G. and Vidal-Majder, A. 1985, *Ap. J.*, **296**, 593.
Guhathakurta, P. 1991, in Clusters and Superclusters of Galaxies, NATO Advanced Study Institute, Institute of Astronomy, Cambridge.
Gunn, J. E. and Peterson, B. A. 1965, *Ap. J.*, **142**, 1633.
Guth, A. H. 1981, *Phys. Rev.*, **D23**, 347.
Gwinn, C. R., Taylor, J. H., Weisberg, J. M. and Rawlings, L. A. 1986, *Astron. J.*, **91**, 338.
Hammer, F. 1991, *Ap. J.*, **383**, 66.
Hammer, F. and Rigaut, F. 1989, *Astr. Ap.*, **226**, 45.
Harari, H. 1989, *Phys. Lett.*, **B216**, 413.
Haud, U. 1992, *M.N.R.A.S.*, **257**, 707.

Hayashi, C. 1950, *Prog. Theor. Phys.*, **5**, 224.

Hayes, K. G. et al 1982, *Phys. Rev.*, **D25**, 2869.

Heavens, A. F. 1991, *M.N.R.A.S.*, **251**, 267.

Heiles, C. 1991, in IAU Symposium No.144, The Interstellar Disk-Halo Connection in Galaxies, ed. Bloemen, H., Kluwer, Dordrecht, p. 433.

Henderson, A. P., Jackson, P. D. and Kerr, F. J. 1982, *Ap. J.*, **263**, 116.

Henry, R. C. 1991, *Ann. Rev. Astr. Ap.*, **29**, 89.

Henry, R. C. and Feldman, P. D. 1981, *Phys. Rev. Lett.*, **47**, 618.

Hill, J. K. 1974, *Astr. Ap.*, **34**, 431.

Hogan, C. J. 1992, *Nature*, **359**, 40.

Hogan, C. J. 1993, *Ap. J.*, **403**, 445.

Hogan, C. J., Kaiser, N. and Rees, M. J. 1982, in The Big Bang and Element Creation, ed. Lynden-Bell, D., *Phil. Trans. Roy. Soc. A.* **307**, 97.

Hogan, C. J. and Weymann, R. J. 1987, *M.N.R.A.S.*, **225**, 1P.

Holberg, J. B. 1993, to be published.

Holberg, J. B. and Barber, H. B. 1985, *Ap. J.*, **292**, 16.

Holzschuh, E. 1992, *Rep. Prog. Phys.*, **55**, 1035.

Hoyle, F. 1948, *M.N.R.A.S.*, **108**, 372.

Hoyle, F. and Tayler, R. J. 1964, *Nature*, **203**, 1108.

Huber, K. P. and Herzberg, G. 1979, Molecular Spectra and Molecular Structure IV. Constants of Diatomic Molecules, New York, van Nostrand Reinhold .

Hughes, J. P. 1989, *Ap. J.*, **337**, 21.

Hughes, J. P. et al 1988, *Ap. J.*, **327**, 615.

Hughes, J. P. and Tanaka, K. 1992, *Ap. J.*, **398**, 62.

Hunstead, R. W., Pettini, M. and Fletcher, A. B. 1990, *Ap. J.*, **356**, 23.

Ikeuchi, S., Murakami, I. and Rees, M. J. 1989, *Pub. Astr. Soc. J.*, **41**, 1095.

Ikeuchi, S. and Turner, E. L. 1991, *Ap. J.*, **381**, L1.

Iwan, D. 1980, *Ap. J.*, **239**, 316.

Jacoby, G. H. et al 1992, *Pub. A.S.P.*, **104**, 599.

Jeans, J. H. 1922, *M.N.R.A.S.*, **82**, 122.

Jenkins, E. B. and Ostriker, J. P. 1991, *Ap. J.*, **376**, 33.

Joblin, C., Leger, A. and Martin, P. 1992, *Ap. J.*, **393**, L79.

Jura, M. 1974, *Ap. J.*, **191**, 375.

Kaiser, N. 1984, *Ap. J.*, **282**, 374.
Kaiser, N. et al 1991, *M.N.R.A.S.*, **252**, 1.
Kapteyn, J. C. 1922, *Ap. J.*, **55**, 302.
Katz, N. 1991, *Ap. J.*, **368**, 325.
Katz, N. and Gunn, J. E. 1991, *Ap. J.*, **377**, 365.
Keh, S. et al 1988, *Phys. Lett.*, **B212**, 123.
Kellermann, K. I. 1993, *Nature*, **361**, 134.
Kenney, J. D. P. 1990, in The Interstellar Medium in Galaxies, ed. Thronson Jr., H. A. and Shull, J. M., Kluwer, Dordrecht, p. 151.
Kennicutt, R. C., Edgar, B. K. and Hodge, P. W. 1989, *Ap. J.*, **337**, 761.
Kent, S. M. and Gunn, J. E. 1982, *Astron. J.*, **87**, 945.
Keppel, J. W., Dettmar, R. J., Gallagher, J. S. and Roberts, M. S. 1991, *Ap. J.*, **374**, 507.
Kimble, R., Bowyer, S. and Jakobsen, P. 1981, *Phys. Rev. Lett.*, **46**, 80.
Kimble, R. et al 1993, *Ap. J.*, **404**, 63.
Kolb, E. W. and Turner, M. S. 1989, *Phys. Rev. Lett.*, **62**, 509.
Kolb, E. W. and Turner, M. S. 1990, The Early Universe, Addison Wesley , Redwood City.
Koo, B. C., Heiles, C. and Reach, W. T. 1992, *Ap. J.*, **390**, 108.
Krauss, L. M., Romanelli, P., Schramm, D. and Lehrer, R. 1992, *Nucl. Phys.*, **B380**, 507.
Kuijken, K. 1991, *Ap. J.*, **376**, 467.
Kuijken, K. and Gilmore, G. 1991, *Ap. J.*, **367**, L9. (KG)
Kuijken, K. and Tremaine, S. 1993, *Ap. J.*, , in press.
Kulkarni, S. R., Blitz, L. and Heiles, C. 1982, *Ap. J.*, **259**, L59.
Kulkarni, S. R. and Heiles, C. 1987, in Interstellar Processes ed. Hollenbach, D. J. and Thronson Jr., H. A., Reidel, Dordrecht, p.87.
Kutyrev, A. S. and Reynolds, R. J. 1989, *Ap. J.*, **344**, L9.
Lambas, D. G., Maddox, S. J. and Loveday, J. 1992, *M.N.R.A.S.*, **258**, 404.
Landau, L. and Lifshitz, E. 1958, Statistical Physics, Pergamon, Oxford.
Langer, W. 1976, *Ap. J.*, **206**, 699.
Lee Po and Weissler, G. L. 1952, *Ap. J.*, **115**, 570.
Lepp, S. and Dalgarno A. 1988, *Ap. J.*, **355**, 769.

Lequeux, J., Peimbert, M., Rayo, J. F., Serrano, A. and Torres-Peimbert, S. 1979, *Astr. Ap.*, **80**, 155.
Lim, C. S. and Marciano, W. J. 1988, *Phys. Rev.*, **D37**, 1368.
Lindblad, B. 1926, *Uppsala Medd.*, **30**, 11.
Linde, A. D. 1990, Particle Physics and Inflationary Cosmology, Harwood Academic Publishers.
Linsky, J. L. et al 1993, *Ap. J.*, **402**, 694.
Longair, M. S. and Sunyaev, R. A. 1972, *Sov. Phys. Usp.*, **14**, 569.
Longo, R., Stalio, R., Polidan, R. S. and Rossi, L. 1989, *Ap. J.*, **339**, 474.
Lowenthal, J. D., Hogan, C. J., Leach, R. W., Schmidt, G. D. and Foltz, C. B. 1990, *Ap. J.*, **357**, 3.
Lowenthal, J. D. et al 1991, *Ap. J.*, **377**, L73.
Lu, L., Wolfe, A. M. and Turnshek, D. A., 1991, *Ap. J.*, **367**, 19.
Lynden-Bell, D. and Gilmore, G. eds. 1990, Baryonic Dark Matter, Kluwer, Dordrecht.
Lynds, R. and Petrosian, V. 1986, *Bull. Amer. Astr. Soc*, **18**, 1014.
Lyne, A. G., Manchester, R. N. and Taylor, J. H. 1985, *M.N.R.A.S.*, **213**, 613.
Lyttleton, R. A. 1958, *M.N.R.A.S.*, **118**, 551.
Maalampi, J. and Roos, M. 1991, *Phys. Lett.*, **B263**, 437.
Madau, P. 1991, *Ap. J.*, **376**, L33.
Madau, P. 1992, *Ap. J.*, **389**, L1.
Madau, P. and Meiksin, A. 1991, *Ap. J.*, **374**, 6.
Maloney, P. 1989, in The Interstellar Medium in External Galaxies ed. Hollenbach, D. J. and Thronson Jr., H. A., NASA, p. 1.
Maloney, P. 1992 private communication.
Maloney, P. 1993, Ap. J. in press.
Marciano, W. and Sanda, A. I. 1977, *Phys. Lett.*, **B67**, 303.
Martin, C. and Bowyer, S. 1989, *Ap. J.*, **338**, 677.
Martin, C. and Bowyer, S. 1990, *Ap. J.*, **350**, 242.
Martin, P. G. 1988, *Ap. J. Suppl.*, **66**, 125.
Marx, G. and Szalay, A. S. 1972, in Neutrino 72 conf. proceedings, Technoinform, Budapest, p. 191.
Mather, J. C. et al 1990, *Ap. J.*, **354**, L37.
Mather, J. C. et al 1993, *Ap. J.*, , in press.
Mathews, G. J., Schramm, D. N. and Meyer, B. S. 1993, *Ap. J.*, **404**, 476.

Mathis, J. S. 1986, *Ap. J.*, **301**, 423.
McCrea, W. H. and Milne, E. A. 1934, *Quart. J. Math.*, **5**, 73.
McKee, C. F. and Ostriker, J. P. 1977, *Ap. J.*, **218**, 418.
Meiksin, A. and Madau, P. 1993, *Ap. J.*, , in press.
Melott, A. L. 1982a, *Phys. Rev. Lett.*, **48**, 894.
Melott, A. L. 1982b, *Nature*, **296**, 721.
Melott, A. L. 1984, *Sov. Astr.*, **28**, 478.
Melott, A. L. and Sciama, D. W. 1981, *Phys. Rev. Lett.*, **46**, 1369.
Melott, A. L., McKay, D. W. and Ralston, J. P. 1988, *Ap. J.*, **324**, L43.
Merrifield, M. R. 1992, *Astron. J.*, **103**, 1552.
Merritt, D. 1987, *Ap. J.*, **313**, 121.
Meszaros, P. 1973, *Ap. J.*, **185**, L41.
Milliard, B., Donas, J., Laget, M., Armand, C. and Vuillemin, A. 1992, *Astr. Ap.*, **257**, 24.
Minakata, H. and Nunokowa, H. 1990, *Phys. Rev.*, **D41**, 2976.
Miralda-Escudé, J. 1993, *Ap. J.*, **497**, 509.
Miralda-Escudé, J. and Ostriker, J. P. 1990, *Ap. J.*, **350**, 1.
Miralda-Escudé, J. and Ostriker, J. P. 1992, *Ap. J.*, **392**, 15.
Mohapatra, R. N. and Pal, P. B. 1991, Massive Neutrinos in Physics and Astrophysics, World Scientific, Singapore.
Moller, P. and Kjaergaard, P. 1992, *Astr. Ap.*, **258**, 234.
Mori, M. et al 1992, *Phys. Lett.*, **B289**, 463.
Morris, S. L., Weymann, R. J., Savage, B. D. and Gilliland, R. L. 1991, *Ap. J.*, **377**, L21.
Munch, G. and Pitz, E. 1990, in IAU Symposium No.139, The Galactic and Extragalactic Background Radiation, ed. Bowyer S. and Leinert, C., Kluwer, Dordrecht, p. 193.
Murdoch, H. S., Hunstead, R. W., Pettini, M. and Blades, J. C. 1986, *Ap. J.*, **309**, 19.
Murthy, J., Henry, R. C. and Holberg, J. B. 1991, *Ap. J.*, **383**, 198.
Nakagawa, M., Okonogi, H., Sakata, S. and Toyoda, A. 1963, *Prog. Theor. Phys.*, **30**, 727.
Narlikar, J. V. and Padmanabhan, T. 1991, *Ann. Rev. Astr. Ap.*, **29**, 325.
Nasel'skij, P. D., Novikov, I. D. and Reznitsky, L. I. 1987, *Sov. Astr.*, **30**, 364.

Nasel'skij, P. D. and Polnarev, A. G. 1987, *Sov. Astr. Lett.*, **13**, 67.
Neufeld, D. A. 1990, *Ap. J.*, **350**, 216.
Nilles, H. P. 1984, *Phys. Rep.*, **110**, 1.
Nordgren, T., Cordes, J. and Terzian, Y. 1992, *Astron. J.*, **104**, 1465.
Norman, C. A. 1991, in IAU Symposium No.144, The Interstellar Disk–Halo Connection in Galaxies, ed. Bloemen, H., Kluwer, Dordrecht p. 337.
Norman, C. A. and Ikeuchi, S. 1989, *Ap. J.*, **345**, 372.
Nousek, J. A., Fried, P. M., Sanders, W. T. and Kraushaar, W. L. 1982, *Ap. J.*, **258**, 83.
Nussinov, S. and Rephaeli, Y. 1987, *Phys. Rev.*, **D36**, 2278.
O'Donnell, E. J. and Watson, W. D. 1974, *Ap. J.*, **191**, 89.
Oegerle, W. R., Fitchett, M. J. and Danly, L. eds. 1990, Clusters of Galaxies, Space Telescope Science Institute, Symposium Series 4, Cambridge University Press.
Oegerle, W. R., Fitchett, M. J., Hill, J. M. and Hintzen, P. 1991, *Ap. J.*, **376**, 46.
Olive, K. A., Schramm, D. N. and Steigman, G. 1981, *Nucl. Phys.*, **B180**, 497.
Onaka, T. and Kodaira, K. 1991, *Ap. J.*, **379**, 532.
Oort, J. H. 1932, *B.A.N.*, **6**, 249.
Oort, J. H. 1965, in Galactic Structure, ed. Blaauw, A. and Schmidt, M., Chicago.
Osterbrock, D. 1989, Astrophysics of Gaseous Nebulae and Active Galactic Nuclei, University Science Books, Mill Valley CA.
Ostriker, J. P. and Peebles, P. J. E. 1973, *Ap. J.*, **186**, 467.
Ostriker, J. P. and Vishniac, E. T. 1986, *Ap. J.*, **306**, L51.
Overduin, J. M., Wesson, P. S. and Bowyer, S. 1993, *Ap. J.*, **404**, 460.
Paczynski, B. 1986a, *Ap. J.*, **301**, 503.
Paczynski, B. 1986b, *Ap. J.*, **304**, 1.
Pagel, B. E. J. and Simonson, E. A. 1989, *Rev. Mex. Astr. Astrofis.*, **18**, 153.
Pagel, B. E. J. and Kazlauskas, A. 1992, *M.N.R.A.S.*, **256**, 49P.
Pagel, B. E. J., Simonson, E. A., Terlevich, R. J. and Edmunds, M. G. 1992, *M.N.R.A.S.*, **255**, 325.
Pal, P. B. and Wolfenstein, L. 1982, *Phys. Rev.*, **D25**, 766.

Paresce,F. 1984, *Astron. J.*, **89**, 1022.
Payne, H. E., Salpeter, E. E. and Terzian, Y. 1984, *Astron. J.*, **89**, 668.
Peacock, J. A. 1991, *Nature*, **349**, 190.
Peebles, P. J. E. 1966, *Ap. J.*, **146**, 542.
Peebles, P. J. E. 1971, Physical Cosmology, Princeton University Press.
Peebles, P. J. E. 1980, in Physical Cosmology, ed. Balian, R., Audouze, J. and Schramm, D. N., North-Holland, Amsterdam p. 265.
Peebles, P. J. E. 1982, *Ap. J.*, **258**, 415.
Peebles, P. J. E. 1986, *Nature*, **321**, 27.
Peebles, P. J. E. 1987, *Ap. J.*, **315**, L51.
Penzias, A. A. and Wilson, R. W. 1965, *Ap. J.*, **142**, 419.
Persic, M. and Salucci, P. 1988, *M.N.R.A.S.*, **234**, 131.
Persic, M. and Salucci, P. 1990a, *M.N.R.A.S.*, **245**, 577.
Persic, M. and Salucci, P. 1990b, *M.N.R.A.S.*, **247**, 349.
Persic, M. and Salucci, P. 1992a, *M.N.R.A.S.*, **258**, 14P.
Persic, M. and Salucci, P. 1992b, in The Distribution of Matter in the Universe, eds Mamon, G. and Gerbal, D. (Paris: Editions de l'Observatoire), in press.
Petcov, S. T. 1977a, *Sov. J. Nucl. Phys.* , **25**, 340.
Petcov, S. T. 1977b, *Sov. J. Nucl. Phys.* , **25**, 698(E).
Pettini, M., Boksenberg, A. and Hunstead, R. W. 1990, *Ap. J.*, **348**, 48.
Pettini, M., Hunstead, R. W., Smith, L. J. and Mar, D. P. 1990, *M.N.R.A.S.*, **246**, 545.
Pierce, M. J., Ressler, M. E. and Shure, M. S. 1992, *Ap. J.*, **390**, L45.
Prasad, S. and Tarafdar, S. P. 1983, *Ap. J.*, **267**, 603.
Press, W. H., Rybicki, G. B. and Hewitt, J. N. 1992a, *Ap. J.*, **385**, 404.
Press, W. H., Rybicki, G. B. and Hewitt, J. N. 1992b, *Ap. J.*, **385**, 416.
Proffitt, C. R. and Michaud, G. 1991, *Ap. J.*, **371**, 584.
Proffitt, C. R. and Vandenberg, D. A. 1991, *Ap. J. Suppl.*, **77**, 473.
Raffelt, G. G. 1990a, *Phys. Rep.*, **198**, 1.
Raffelt, G. G. 1990b, *Ap. J.*, **365**, 559.

Raffelt, G. G. and Weiss, A. 1992, *Astr. Ap.*, **264**, 536.
Ralston, P., McKay, D. and Melott, A. L. 1988, *Phys. Lett.*, **B202**, 40.
Rand, R. J., Kulkarni, S. R. and Hester, J. J. 1990a, *Ap. J.*, **352**, L1. (RKH)
Rand, R. J., Kulkarni, S. R. and Hester, J. J. 1990b, *Ap. J.*, **362**, L35.
Rand, R. J., Kulkarni, S. R. and Hester, J. J. 1992, *Ap. J.*, **396**, 97.
Rauch, M. et al 1992, *Ap. J.*, **390**, 387.
Rauch, M. et al 1993, *M.N.R.A.S.*, **260**, 589.
Readhead, A. C. S. and Duffett-Smith, P. J. 1975, *Astr. Ap.*, **42**, 151.
Readhead, A. C. S. et al 1989, *Ap. J.*, **346**, 566.
Rees, M. J. 1990 private communication.
Reimers, D. et al 1992, *Nature*, **360**, 561.
Renzini, A. 1991, in Observational Tests of Cosmological Inflation, ed. Shanks, T. et al, Kluwer, Dordrecht, p. 131.
Rephaeli, Y. and Szalay, A. S. 1981, *Phys. Lett.*, **B106**, 73.
Reynolds, R. J. 1984, *Ap. J.*, **282**, 191.
Reynolds, R. J. 1987, *Ap. J.*, **323**, 118.
Reynolds, R. J. 1989a, *Ap. J.*, **339**, L29.
Reynolds, R. J. 1989b, *Ap. J.*, **345**, 811.
Reynolds, R. J. 1989c, in IAU Symposium No.139, Galactic and Extragalactic Background Radiation, ed. Bowyer, S. and Leinert, C., Reidel, Dordrecht, p. 157.
Reynolds, R. J. 1990a, *Ap. J.*, **348**, 153.
Reynolds, R. J. 1990b, *Ap. J.*, **349**, L17.
Reynolds, R. J. 1991, in IAU Symposium No.144, The Interstellar Disk-Halo Connection in Galaxies, ed. Bloemen, H., Kluwer, Dordrecht, p. 67.
Reynolds, R. J. 1992, *Ap. J.*, **392**, L35.
Reynolds, R. J. and Cox, D. P. 1992, *Ap. J.*, **400**, L33.
Reynolds, R. J., Roesler, F. L. and Scherb, F. 1977, *Ap. J.*, **211**, 115.
Reynolds, R. J., Magee, K., Roesler, F. L., Scherb, F. and Harlander J. 1986, *Ap. J.*, **309**, L9.
Rhee, G. 1991, *Nature*, **350**, 211.
Richer, H. B. and Fahlman, G. G. 1992, *Nature*, **358**, 383.

Richstone, D. et al 1992, *Ap. J.*, **388**, 354.
Rindler, W. 1956, *M.N.R.A.S.*, **116**, 662.
Rood, H. J. 1982, *Rep. Prog. Phys.*, **44**, 1077.
Rosen, S. P. and Gelb, J. M. 1989, *Phys. Rev.*, **D39**, 3190.
Roulet, E. and Tommasini, D. 1991, *Phys. Lett.*, **B256**, 218.
Rowan-Robinson, M. 1985, The Cosmological Distance Ladder, Freeman, New York.
Rubin, V. C. 1979, *Comm. Astrophys.*, **8**, 79.
Rubin, V. C. 1991, in After the First Three Minutes: AIP Conference Proceedings 222, eds Holt, S. S., Bennet, C. L. and Trimble, V. , AIP New York.
Sachs, R. K. and Wolfe, A. M. 1967, *Ap. J.*, **147**, 73.
Sackett, P. D. and Sparke, L. S. 1990, *Ap. J.*, **361**, 408.
Sackett, P. D. 1991, in Warped Disks and Inclined Rings around Galaxies, ed. Casertano, S., Sackett, P. D. and Briggs, F., Cambridge University Press, p. 73.
Salati, P. and Wallet, J. C. 1984, *Phys. Lett.*, **B144**, 61.
Salucci, P. and Frenk, C. S. 1989, *M.N.R.A.S.*, **237**, 247.
Salucci, P. and Sciama, D. W. 1990, *M.N.R.A.S.*, **244**, 9P.
Salucci, P. and Sciama, D. W. 1991, *Astr. Ap.*, **243**, 341.
Salucci, P., Frenk, C. S. and Persic, M. 1993, *M.N.R.A.S.*, , in press.
Salucci, P., Persic, M. and Borgani, S. 1993, *Ap. J.*, **405**, 459.
Sancisi, R. and Allen, R. J. 1979, *Astr. Ap.*, **74**, 73.
Sandage, A. 1993a, *Ap. J.*, **402**, 3.
Sandage, A. 1993b, *Ap. J.*, **404**, 492.
Sandage, A., Saha, A., Tammann, G., Panagia, N. and Macchetto, F. 1992, *Ap. J.*, **401**, L7.
Sarazin, C. 1988, X-Ray Emission from Clusters of Galaxies, Cambridge University Press.
Sargent, W. L., Steidel, C. C. and Boksenberg, A. 1989, *Ap. J. Suppl.*, **69**, 703.
Sargent, W. L., Young, P. J., Boksenberg, A. and Tytler, D. 1980, *Ap. J. Suppl.*, **42**, 41.
Sarkar, S. 1991, in Observational Tests of Cosmological Inflation, ed. Shanks, T. et al, Kluwer, Dordrecht, p. 91.
Schechter, J. and Valle, J. W. F. 1981, *Phys. Rev.*, **D24**, 1883.
Schmidt, B. P., Kirshner, R. P. and Eastman, R. G. 1992, *Ap. J.*, **395**, 366.

Schmidt, M. 1965, *Ap. J.*, **141**, 1295.
Schmidt, M. 1966, *Ap. J.*, **162**, 371.
Schneider, P., Ehlers, J., and Falco, E. E. 1992, Gravitational Lenses, Springer, Berlin.
Schneps, J. 1991, in Third International Workshop on Neutrino Telescopes, ed. Balda-Ceolin, M., Venice.
Schneps, J. 1992, in Neutrino 92, Granada, Spain.
Schweizer, F., Whitmore, B. C. and Rubin, V. C. 1983, *Ap. J.*, **88**, 909.
Sciama, D. W. 1982a, *Phys. Lett.*, **B112**, 211.
Sciama, D. W. 1982b, *Phys. Lett.*, **B114**, 19.
Sciama, D. W. 1982c, *M.N.R.A.S.*, **198**, 1P.
Sciama, D. W. 1982d, *Phys. Lett.*, **B118**, 327.
Sciama, D. W. 1984, in The Big Bang and Georges Lemaitre, ed. Berger, A., Reidel, Dordrecht, p. 31.
Sciama, D. W. 1988, *M.N.R.A.S.*, **230**, 13P.
Sciama, D. W. 1990a, *Ap. J.*, **364**, 549.
Sciama, D. W. 1990b, *Phys. Rev. Lett.*, **65**, 2839.
Sciama, D. W. 1990c, *Nature*, **346**, 40.
Sciama, D. W. 1990d, *Nature*, **348**, 617.
Sciama, D. W. 1991a, *Ap. J.*, **367**, L39.
Sciama, D. W. 1991b, in The Early Observable Universe from Diffuse Backgrounds, ed. Rocca-Volmerange, B., Deharveng, J. M. and Tran Thanh Van, J. , Edition Frontieres, p.127.
Sciama, D. W. 1992, *Int. Journ. of Mod. Phys. D*, **1**, 161.
Sciama, D. W. 1993a, *Ap. J.*, **415**, L31.
Sciama, D. W. 1993b, *Ap. J.*, **409**, L25.
Sciama, D. W. and Melott, A. L. 1982, *Phys. Rev.*, **D25**, 2214.
Sciama, D. W. and Rees, M. J. 1966, *Nature*, **211**, 1283.
Sciama, D. W. and Salucci, P. 1990, *M.N.R.A.S.*, **247**, 506.
Sciama, D. W., Salucci, P. and Persic, M. 1992, *Nature*, **358**, 718.
Sciama, D. W., Persic, M. and Salucci, P. 1993, *Pub. A.S.P.*, **105**, 102.
Scott, D. 1993, *M.N.R.A.S.*, **263**, 903.
Scott, D., Rees, M. J. and Sciama, D. W. 1991, *Astr. Ap.*, **250**, 295.
Sembach, K. R., Savage, B. D. and Massa, D. 1991, *Ap. J.*, **372**, 81.
Shapiro, P. R. and Giroux, M. L. 1987, *Ap. J.*, **321**, L107.

Shapiro, P. R. and Giroux, M. L. 1989, in The Epoch of Galaxy Formation, ed. Frenk, C. S. et al, Kluwer, Dordrecht p. 153.
Shapiro, S. L., Teukolsky, S. A. and Wassermann, I. 1980, *Phys. Rev. Lett.,* **45**, 669.
Shaver, P. A. 1976, *Astr. Ap.,* **49**, 149.
Shaver, P. A., Pedlar, A. and Davies, R. D. 1976, *M.N.R.A.S.,* **177**, 45.
Shaver, P. A., Wampler, E. J. and Wolfe, A. M. 1991 eds., Quasar Absorption Lines, ESO Scientific Report No. 9.
Shi, X., Schramm, D. N. and Bahcall, J. N. 1992, *Phys. Rev. Lett.,* **69**, 717.
Shipman, H. L. and Cowsik, R. 1981, *Ap. J.,* **247**, L111.
Silk, J. 1992, in IAU Symposium No. 149, Stellar Populations, ed. Barbuy, B. and Renzini, A., Kluwer, Dordrecht.
Silk, J. and Sunyaev, R. A. 1976, *Nature,* **260**, 508.
Silk, J. and Werner, M. W. 1970, *Ap. J.,* **158**, 185.
Silk, J., Wyse, R. and Shields, G. A. 1987, *Ap. J.,* **322**, L59.
Sivan, J. P., Stasinska, G. and Lequeux, J. 1986, *Astr. Ap.,* **158**, 279.
Slavin, J. D. 1989, *Ap. J.,* **346**, 718.
Smirnov, Y. N. 1965, *Sov. Astr.,* **8**, 864.
Smith, H. E., Cohen, R. D., Burns, J. E., Moore, D. J. and Uchida, B. A. 1989, *Ap. J.,* **347**, 87.
Smith, M. S., Kawano, L. H. and Malaney, R. A. 1993, *Ap. J. Suppl.,* **85**, 219.
Smith, P. F. and Lewin, J. D. 1990, *Phys. Rep.,* **187**, 204.
Smoot, G. F. et al 1992, *Ap. J.,* **396**, L1.
Songaila, A., Bryant, W. and Cowie, L.L. 1989, *Ap. J.,* **345**, L71.
Songaila, A., Cowie, L. L. and Lilly, S. J. 1990, *Ap. J.,* **348**, 371.
Sorrell, W. H. 1992, *Comm. Astrophys.,* **16**, 123.
Soucail, G. and Fort, B. 1991, *Astr. Ap.,* **243**, 23.
Soucail, G., Fort, B., Mellier, Y. and Picat, J.-P 1987, *Astr. Ap.,* **172**, 414.
Sparke, L. S. and Casertano, S. 1988, *M.N.R.A.S.,* **234**, 873.
Spitzer, L. 1978, Physical Processes in the Interstellar Medium, Wiley, New York.
Spitzer, L. and Tomasko, M. G. 1968, *Ap. J.,* **152**, 971.
Spitzer, L. and Fitzpatrick, E. L. 1992, *Ap. J.,* **391**, L41.
Spitzer, L. and Fitzpatrick, E. L. 1993, *Ap. J.,* **409**, 299.

Stalio, R., Bowyer, S., Sciama, D. W., and Gimenez, A. 1992, *Adv. in Space Research*, in press.

Stecker, F. W. 1980, *Phys. Rev. Lett.*, **45**, 1460.

Steidel, C. C. and Sargent, W. L. W. 1987, *Ap. J.*, **318**, L11.

Steidel, C. C. and Sargent, W. L. W. 1989, *Ap. J.*, **343**, L33.

Stocke, J. T., Case, J., Donahue, M., Shull, J. M. and Snow, T. P. 1991, *Ap. J.*, **374**, 72.

Subramanian, K. 1988, *M.N.R.A.S.*, **234**, 459.

Sunyaev, R. A. 1969, *Astrophys. Lett.*, **3**, 33.

Tammann, G. A. 1992, Phys. Script. to be published.

Tarafdar, S. P. 1991, *M.N.R.A.S.*, **252**, 55P.

Tayler, R. J. 1990, *Quart. J.R.A.S.*, **31**, 371.

Tegmark, M., Silk, J. and Evrard, A. 1993, to be published.

The, L. S. and White, S. M. 1986, *Astron. J.*, **92**, 1248.

Thomas, P. A. and Fabian, A. C. 1990, *M.N.R.A.S.*, **246**, 156.

Tielens, A. G. G. M. and Hollenbach, D. 1985a, *Ap. J.*, **291**, 722.

Tielens, A. G. G. M. and Hollenbach, D. 1985b, *Ap. J.*, **291**, 747.

Tommasini, D. 1993, Phys. Lett. B, in press.

Torres-Peimbert, S., Peimbert, M. and Fierro, J. 1989, *Ap. J.*, **345**, 186.

Tremaine, S. and Gunn, J. E. 1979, *Phys. Rev. Lett.*, **42**, 407.

Trimble, V. 1987, *Ann. Rev. Astr. Ap.*, **25**, 425.

Tully, R. B. 1989, Nearby Galaxy Catalogue, Cambridge University Press.

Turner, E. L., Ostriker, J. P. and Gott, J. R. 1984, *Ap. J.*, **284**, 1.

Turner, M. S. 1990, *Phys. Rep.*, **197**, 67.

Urban, M. et al 1992, *Phys. Lett.*, **B293**, 149.

van Albada T. S., Bahcall, J. N., Begeman, K. and Sancisi, R. 1985, *Ap. J.*, **295**, 305.

Vandenberg, D. A. 1992, *Ap. J.*, **391**, 685.

van den Bergh, S. 1989, *Astr. Ap. Rev.*, **1**, 111.

van den Bergh, S. 1992a, *Science*, **258**, 421.

van den Bergh, S. 1992b, *Pub. A.S.P.*, **104**, 861.

van der Kruit, P. C. 1984, *Astr. Ap.*, **140**, 470.

van der Kruit, P. C. 1987, *Astr. Ap.*, **173**, 59.

van der Marel, R. P. 1991, *M.N.R.A.S.*, **248**, 515.

van Dieshoeck, E. F. and Black, J. H. 1988, *Ap. J.*, **334**, 71.

Vangioni-Flam, E., Casse, M., Audouze, J. and Tran Thanh Van, J. eds., 1990, Astrophysical Ages and Dating Methods, Editions Frontieres, Cedex.

van Gorkom, J. et al 1989 in preparation.

Verstraete, L., Leger, A., d'Hendecourt B., Dutuit, O. and Defourneau, D. 1990, *Ap. J.,* **237**, 436.

Vishniac, E. T. 1987, *Ap. J.,* **322**, 597.

Wagoner, R. V., Fowler, W. A. and Hoyle, F. 1967, *Ap. J.,* **148**, 3.

Walker, A. R. 1992, *Ap. J.,* **390**, L81.

Walker, T. P., Steigman, G., Schramm, D. N., Olive, K. A. and Kang, H. 1991, *Ap. J.,* **376**, 51.

Walterbos, R. A. M. 1991, in IAU Symposium No.144, The Interstellar Disk-Halo Connection, ed. Bloemen, H., Kluwer, Dordrecht.

Warren, M. S., Quinn, P. J., Salmon, J. K. and Zurek, W. H. 1992, *Ap. J.,* **399**, 405.

Watson, R. A. et al 1992, *Nature,* **357**, 660.

Webb, J. K., Barcons, X., Carswell, R. F. and Parnell, H. C. 1992, *M.N.R.A.S.,* **255**, 319.

Weinberg, S. 1972, Gravitation and Cosmology, Wiley, New York.

Weinberg, S. 1989, *Rev. Mod. Phys.,* **61**, 1.

Weisberg, J. M., Rankin, J. M. and Boriakoff, V. 1980, *Astr. Ap.,* **88**, 84.

Werner, M. W., Silk, J. and Rees, M. J. 1970, *Ap. J.,* **161**, 965.

White, D. A., Fabian, A. C., Johnstone, R. M., Mushotzky, R. F. and Arnaud, K. A. 1991, *M.N.R.A.S.,* **252**, 72.

White, S. D. and Frenk, C. S. 1991, *Ap. J.,* **379**, 52.

Winter, K., ed., 1991, Neutrino Physics, Cambridge University Press..

Winter, K. 1992, in Fourth International Workshop on Neutrino Telescopes, ed. Balda-Ceolin, M., Venice.

Wolfe, A. M. 1991, in Quasar Absorption Lines, ed. Shaver, P. A., Wampler, E. J. and Wolfe, A. M., ESO Scientific Report No. 9.

Wolfe, A. M., Turnshek, D. A., Smith, H. E. and Cohen, R. D. 1986, *Ap. J. Suppl.,* **61**, 249.

Wolfe, A. M., Turnshek, D. A., Lanzetta, K. M. and Oke, J. B. 1992, *Ap. J.,* **385**, 151.

Wright, E. L. 1992, *Ap. J.*, **391**, 34.
Wright, E. L. et al 1991, *Ap. J.*, **381**, 200.
Yanagida, T. 1978, *Prog. Theor. Phys.*, **135**, 66.
York, D. G. et al 1983, *Ap. J.*, **266**, L55.
Zmuidzines, J., Betz, A. L. and Goldhaber, D. M. 1986, *Ap. J.*, **307**, L75.
Zwicky, F. 1933, *Helv. Phys. Acta*, **6**, 110.

Subject index

Printed in the United States
By Bookmasters